Learning Commons Treasury

Edited by:

David V. Loertscher

Elizabeth "Betty" Marcoux

DISCARD

Teacher Librarian Press

2010

Copyright © 2010 Teacher Librarian Press, an imprint of E L Kurdyla Publishing LLC

All rights reserved. No part of this publication may be reproduced in any form or by any means, except for reasonable and brief quotes for reviews, professional presentations, and educational purposes only, without prior written permission of the publisher.

Content current through April 20, 2010

All rights reserved

E L Kurdyla Publishing LLC
PO Box 958
Bowie, MD 20718-0958

Distributed by:

LMC Source
P.O. Box 131266
Spring TX 77393
800-873-3043
email: sales@lmcsource.com
Website: http://lmcsource.com

ISBN: 978-1-61751-000-7

Contents

Preface v

Part I: Foundation 1

1. The Time is Now: Transform Your School Library into a Learning Commons by Carol Koechlin, Sandi Zwaan, and David V. Loertscher 3

Part II: Learning Commons Examples 11

2. From Library to Learning Commons: a Metamorphosis by Valerie Diggs with Editorial Comments by David Loertscher 13
3. Concord-Carlisle Transitions to a Learning Commons by Robin Cicchetti 20
4. The Learning Commons is Alive and Well in New Zealand by Peggy Stedman and Greg Carroll 27
5. From Book Museum to Learning Commons: Riding the Transformation Train by Christina A. Bentheim 31

Part III: Curriculum and the Learning Commons 35

6. Curriculum, the Library/Learning Commons and Teacher Librarians: Myths and Realities in the Second Decade by David Loertscher 37
7. The Library is the Place: Knowledge and Thinking, Thinking and Knowledge by Derek Cabrera and Laura Colosi 43
8. The Role of the School Library in the Reading Program by Elizabeth "Betty" Marcoux and David V. Loertscher 49
9. Influencing Positive Change: The Vital Behaviors to Turn Schools Toward Success by Vicki Davis 55
10. Everyone Wins: Differentiation in the School Library by Carol Koechlin and Sandi Zwaan 60
11. Cultivating Curious Minds: Teaching for Innovation Through Open-Inquiry Learning by Jean Sausele Knodt 66
12. Information Literate? Just Turn the Children Loose! By Joy Mounter 73
13. Gifted Readers and Libraries: A Natrual Fit by Rebecca Haslam-Odoardi 77
14. Using the library learning commons to reengage disengaged students and making it a student-friendly place by Cynthia Sargeant and Roger Nevin 81a
15. Rethinking collaboration: transforming Web 2.0 thinking into real time behavior by Sheila Cooper-Simon 81d

Part IV: Technology and the Learning Commons 83

16. Achieving Teaching and Learning Excellence with Technology by Elizabeth "Betty" Marcoux and David V. Loertscher 85
17. Supporting 21st Century Learning Through Google Apps by Roger Nevin. 94
18. The Effect of Web 2.0 on Teaching and Learning by Richard Byrne 98
19. WLANS for the 21st Century Library by Cal Calamari.......................101
20. The Impact of Facebook on our Students by Doug Fodeman and Marje Monroe 104

Part V: Leadership in the Learning Commons 109

21. Librarians and Learning Specialists: Moving From the Margins to the Mainstream of School Leadership by Alison Zmuda and Violet H. Harada 111
22. Information and Technology Literacy by Bill Derry.........................117
23. Technology Leadership: Kelly Czarnecki by the TL Editors 120
24. Three Heads and Better Than One: The Reading Coach, The Classroom Teacher, and the Teacher Librarian, by Christopher Lamb, Winnie Porter and Carol Lopez 122
25. Advanced Contemporary Literacy: An Integrated Approach to Reading by Sharon Swarner 124

Part VI: Assessment in the Learning Commons 127

26. Creating Personal Learning Through Self-Assessment by Jean Donham 129
27. Our Instruction Does Matter: Data Collected from Students' Works Cited Speaks Volumes by Sarah Pointer and Jennifer Alevy 137
28. The Big Think: Reflecting, Reacting, and Realizing Improved Learning by Carol Koechlin and Sandi Zwaan................................139

Index 144

Preface

In 2008, Loertscher, Koechlin and Zwaan challenged school librarians and technology leaders to reinvent the idea of school libraries and computer labs in their book *The New School Learning Commons Where Learners Win*. Noting major shifts in technologies and in the very nature of the new socially networked children and teens, they challenged professionals to do 180 degree thinking.

Instead of inching forward in evolving organizations in the school, the trio recommended a client-side focus to a Learning Commons that had both physical and virtual dimensions. The concept shift has met with some success in the U.S. and Canada.

As of this writing, five publications support the Learning Commons concept:

- Three books:
    - *The New School Learning Commons Where Learners Win* by David V. Loertscher, Carol Koechlin, and Sandi Zwaan (Hi Willow Research & Publishing, 2008)
    - *Building the Learning Commons: A Guide for School Administrators and Learning Leadership Teams: A Whole School Approach to Learning for the Future* by Carol Koechlin, Esther Rosenfeld, and David V. Loertscher (Hi Willow Research & Publishing, 2010)
    - *Learning Commons Treasury*. Edited by David V. Loertscher and Elizabeth "Betty" Marcoux (Teacher Librarian Press, 2010)
- A Wiki
    - http://schoollearningcommons.pbworks.com
- A Periodical
    - *Teacher Librarian*

In the last several years, the editors have invited a number of talented writers, teacher librarians, and researchers to publish articles on the cutting edge of programs leading the theory of having a Learning Commons. This compendium is a collection of articles originally published in *Teacher Librarian*. They provide a range of perspectives and ideas that will inform and encourage the rethinking of what school libraries and computer labs can become. Their original formatting in the periodical has been preserved for this collection so that they can be accessed through the journal as well as in this compendium.

As editors, we extend our appreciation to the writers of these articles for sharing their vision and insight with the profession. Their work is likely to affect many young people across North American and beyond.

Part I:

Foundation

The foundational ideas for the transformation of a school library and computer lab into a learning commons was first set forth in Loertscher, Kechlin, and Zwaan's book: *The New Learning Commons Where Learners Win (2008)*.

For the past half century, school libraries have been a centralized collection of materials and multimedia for the school. It has been *the* place to acquire, store, circulate, and manage a wide array of items and promote the use of those materials throughout the school in support of teaching and learning. The library not only centralized collections but conducted many classes on how to use those materials in support of research projects. School librarians were also dedicated to promote the love of reading to children and teens in an effort to build a life-long reading habit.

Then, as the Internet developed and Google appeared, students and teachers began to bypass their school's library collection in such numbers, that the function of the school library began to be questioned by many.

Likewise, when computer labs began ~~by~~ introducing computers and certain software programs to young people in the era when computers were very expensive and in short supply, they conducted many classes on how to use them in support of research and research projects. However, the more ubiquitous computers and digital devices became, the more students became computer savvy and worked digitally often without the computer lab's assistance. The function of learning computer and software basics came under fire.

This compendium extends an invitation to move way beyond these traditional understandings.

In their foundational article, Koechlin and Zwaan set forth a rationale to change the direction of the traditional school library and computer lab into a client-side organization that emerges as the center of teaching and learning in the school. It becomes a place of collaboration, collaborative contribution and creation; a place owned both physically and virtually by everyone in the school. It is quite a different dream that the model it is designed to replace.

FEATURE ARTICLE

the time is now: transform your school library into a learning commons

IT IS TIME TO CLOSE THE GAP BETWEEN WHAT WE KNOW AND WHAT WE DO IN SCHOOLS. WE KNOW READING CAPACITY IMPROVES STUDENT PERFORMANCE IN ALL FACETS OF LEARNING. WE KNOW THAT TECHNOLOGIES CAN BE EFFECTIVE AND EFFICIENT TOOLS FOR MAKING WORK EASIER, MORE ACCURATE, AND MORE FUN. WE KNOW THAT THERE IS A REAL DISCONNECT BETWEEN WHAT OUR STUDENTS DO WITH TECHNOLOGIES OUTSIDE OF SCHOOL AND WHAT THEY ARE ALLOWED TO DO WITH THEM IN SCHOOL.

We know what the needs of 21st-century learners are, and we are already behind on that schedule. We know that teacher-librarians make a significant contribution to student achievement, and at the same time, we see very little evidence of recognition through programs or staffing. We know that schools that employ collaborative teams for the task of school improvement make a difference, and yet isolation of specialists continues.

A MOMENT IN TIME

A quick tour of Higher High Secondary School finds the teacher-librarian frustrated because yet again another teacher has booked the library only to have access to the Internet:

• The teacher-librarian has read the research and has lots of ideas for facilitating interventions during projects, to really help students build deeper understanding, but the teachers are always too busy. "Just show them where to find the stuff!" they exclaim.

• Down the hall, the teacher technologist sits, head in hand, weary from her daily run of emergency calls from staff and students who get tripped up with techno glitches, lost equipment, and firewalls that deny them access to the data they need. A student who hacked his way onto a social networking site sits outside her door awaiting a reprimand.

• Secluded in the east wing is the student success teacher, recently hired, following a mandate from the district to reach at-risk kids. The dropout rates are too high, but the new hire is running into roadblocks at every turn in her attempts to get needy students released from class.

• The guidance counselor is compiling yet another memo to the Literacy Committee requesting time on the agenda to discuss recent demographic statistics, but the committee has a packed agenda of mandates from administration.

• The arts coordinator sits outside the vice principal's door in panic because she needs an administrator's signature on her grant proposal, and it is due in one hour.

Across the street at Paradise Primary School, the principal is trying to track down the literacy coach because he has money to spend on classroom books, and it must be spent within the next 24 hours:

• The teacher-librarian here is herding one class out and another in for a scheduled book exchange while trying to help a teacher find recent books on energy sources for their next class, but the library budget has been cut, and there are no up-to-date books on energy.

• The vice principal is dealing with a student who brought his iPod to class for the third time this week, and personal portable technology is not allowed at Paradise Primary School.

Do we need to go on?

The images are well focused but wanting. Highly trained specialists are all working at cross purposes, all fighting to get the attention of the classroom teacher and administration, all very dedicated professionals who really desire the best for students. Their earnest efforts result in little or no progress, lots of frustration, inefficient use of time, and students still in need of resources and programs to improve their achievement.

IS THERE A BETTER WAY?

Teaching professionals can paint a better picture, compose a more dynamic

by carol koechlin, sandi zwaan, and david v. loertscher

FIGURE 1

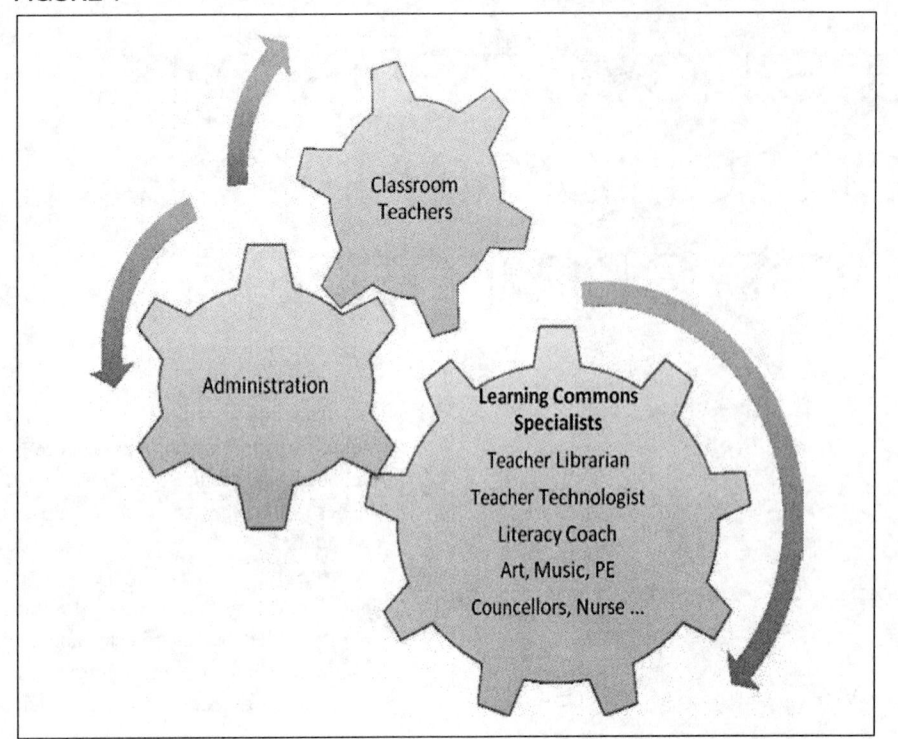

[Figure 1: Three interlocking gears labeled "Classroom Teachers," "Administration," and "Learning Commons Specialists" (Teacher Librarian, Teacher Technologist, Literacy Coach, Art, Music, PE, Councellors, Nurse...)]

"Highly trained specialists are all working at cross purposes, all fighting to get the attention of the classroom teacher and administration, all very dedicated professional who really desire the best for students."

symphony, dance a more relevant step, and write a more powerful story. In the new moment, the cast of players remains the same; the vision is still to advance school improvement; only, the dynamics change to create a synergy and efficacy that is more collaborative, more current, and more successful. Currently, few schools really take advantage of the expertise they have in their teacher-librarian and teacher technologist. These folks hold the key to real-world teaching and learning in this knowledge age. They are highly trained teacher specialists who know how to work with information and technologies to help students learn how to learn in our evolving information places and spaces. They know how to harness collegiality and bring together classroom teachers and the other specialists in a constructive process to design schools to meet the growing complexities of working, playing, and growing in today's world. Similarly we find that other specialists in the school battle at cross purposes for time, budget, and the authority to do what they know needs to be done, and they ask themselves why things aren't working. It's time to stop asking why things aren't working and start asking what we need to do to make it work.

First, we must recognize that we are not doing the best possible job connecting to learners today and get with the program now because we are already a decade behind. Second, schools need to take an honest look at how effectively we are utilizing the school library, computer lab facilities and staff, and the other school specialists. The new question is: How can we utilize our current assets to reorganize, refocus, and reconnect in order to create a learning environment that meets the needs of 21st-century teaching and learning?

Our answer to the current dilemma is to ask teacher-librarians to lead the journey in creating a schoolwide learning commons and challenge administration and staff to invest in this philosophy and build the collaborative learning community necessary to design and sustain the best possible teaching and learning environments. The dynamics of the school learning commons must revolve around learning partnerships that share the vision of creating powerful learning environments combined with the best learning science and a common goal of improving learning and achievement for each and every student.

"It's time to stop asking why things aren't working and start asking what we need to do to make it work."

THE 21ST-CENTURY SCHOOL LEARNING COMMONS

The role and function of the 21st-century school library has evolved as the result of the exponential growth of information, rapid technology advancements, and the challenge to contribute to student achievement. Libraries of the past supplied resources and provided support. Computer labs provided scheduled access and support to technology. Today we need a learning commons, a learning laboratory that is the foundation of all learning in the school rather than a warehouse of information and technologies. The new learning commons focuses on client-centered programs pushing world-class excellence throughout the school. Teacher-librarians in partnerships with teacher technologists (the former tech directors) and other learning specialists must look for ways to capitalize on the rich resources, technologies, spaces, and expertise available in the school library to advance best-practice pedagogy and energize teaching and learning for today's students and teachers. This new mandate is centered on student performance through improved teaching and learning throughout the school.

OUR VISION

Primarily, the school learning commons is the showcase for high-quality teaching and learning—a place to develop and demonstrate exemplary educational practices. It will serve as the professional development center for the entire school—a place to learn, experiment with, assess, and then widely adopt improved instructional programs. It is the keystone of literacy and technological programs of the school and the place where classroom teachers can collaboratively design, build, implement, and assess knowledge-building learning activities.

Much more than a physical space, the library is now also available 24/7/365 as a virtual learning center. Many critical learning activities are happening simultaneously in both the physical and the virtual commons, beginning before school and ending long after the traditional school day is over. One observes the constant flow of individual students, small groups, and large groups through the physical and virtual center each day as personal needs, assignments, and learning activities from regular classrooms are in progress. At the same time, other groups are part of the experimental nature of the commons, including professional development, traditional literacy program, information literacy, emerging literacies, and technology trials. Teachers are coming for individual and group professional development that is planned in concert with their professional learning communities. Parents and other members of the public are serving as volunteers to help make the organization run smoothly. Support staff is maintaining the nuts and bolts of the facility so that it is actually operational. Administrators, specialists, and classroom teachers frequent the center as they plan, implement, and assess the various program components. The learning commons exudes a culture of continuous change and learning. Suddenly, this transformation is recognized in the general educational literature unlike the current practice of ignoring what we do.

For students, the new library is an essential element in their education—a gateway into the vast world of information, exploration, inquiry, learning, interactivity, creativity, and production. It is the center of a world-class education where both individual and collaborative team learning happens. It is the place to learn how to

---

"Many critical learning activities are happening simultaneously in both the physical and the virtual commons, beginning before school and ending long after the traditional school day is over."

Teacher-librarians in partnerships with teacher technologists (the former tech directors) and other learning specialists must look for ways to capitalize on the rich resources, technologies, spaces, and expertise available in the school library to advance best-practice pedagogy and energize teaching and learning for today's students and teachers.

build, manage, and effectively use the Internet and to learn how to manage one's self safely in that space. Again, instead of ignoring us, or just Googling around us, students help build the former library web site into a virtual learning commons.

For teachers, the new library embraces the building of student literacies and inquiry learning as central to their program. It provides an opportunity to share the expertise of other professionals and reignite passion for teaching. When students and teachers work, read, and play with information and ideas in the ever-changing landscape of the learning commons, they, too, experience transformations as they learn and grow. The program elements of the learning commons provide a solid foundation of keystones for 21st-century learning.

LEARNING LITERACIES KEYSTONE

In the past, librarians have promoted literacy and lifetime reading habits for all students, but today, the school's literacy team sees the library as central to planning, implementing, and assessing the literacy effort of the entire school. The literacy team includes administrators, reading coaches, language specialists, classroom teachers, teacher technologists, and the teacher-librarian. This team forms a professional learning community to plan, build, and assess the reading program using evidence-based practices to develop world-class readers who not only know how to read and read to learn, but who are avid readers as well.

The mandate of the learning commons is to prepare students with emerging literacies that will ensure learning for life. As well as reading literacies, the rich environments of the library commons ensure that other necessary student literacies are developed. Students become efficient and effective users of information and evolving technologies.

The exponential growth of information requires that every young person have the skills and abilities needed to be a creative and critical thinker in a world of a thousand voices demanding attention. In order to produce successful learners in the 21st century, schools must

FIGURE 2

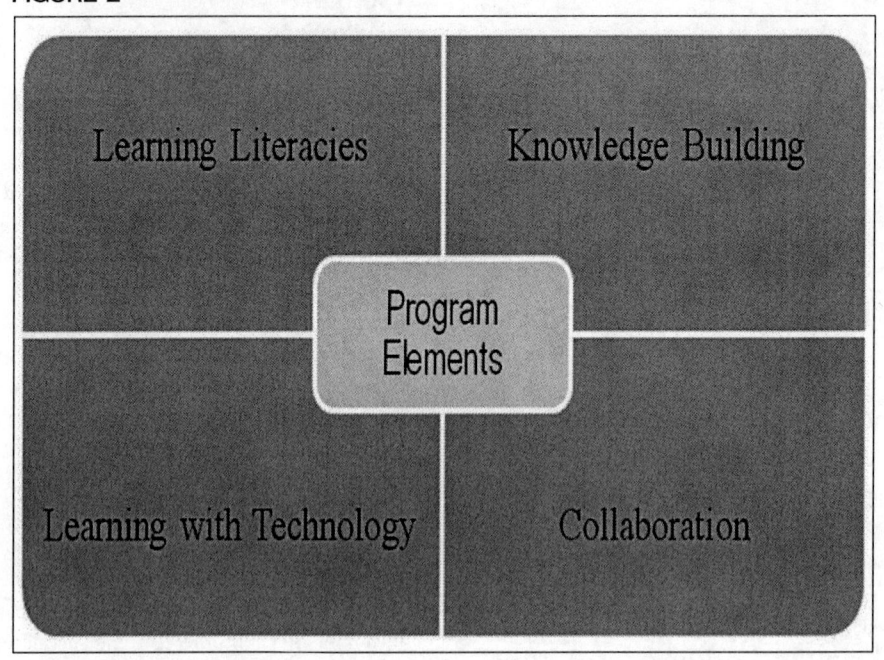

Learning Literacies | Knowledge Building
Program Elements
Learning with Technology | Collaboration

integrate information literacy throughout the whole curriculum. Information literacy has been simplistically defined as the ability to find and use information to meet a personal need. Learning standards developed by the American Association of School Librarians (AASL) in 2007 point to a much more complex intellectual and attitudinal strategy that learners need to develop the information literacy needed to compete in a global society.

Through professional partnerships with teacher-librarians, needed interventions can be effectively designed to infuse information literacy instruction and assessment into all curriculum areas. Through the virtual spaces of the library learning commons, additional support for information literacy will be available to students and teachers around the clock every day.

KNOWLEDGE-BUILDING KEYSTONE

To succeed intellectually, a learner must be a capable and avid reader who is able to follow a process of inquiry. Inquiry skills include beginning with background knowledge, developing questions, finding and evaluating information, reading/viewing/listening to gain understanding, thinking critically, drawing conclusions, answering the questions posed, asking "So what?," sharing, and assessing how well the inquiry has progressed. As learners follow such an inquiry, they develop personal qualities, such as responsibility, positive attitudes toward learning, ethical uses of information and technology, creativity, and self-assessment. Altogether, this leads not only to information-literate students but also to lifelong learners, knowledge builders, and people ready for the challenges of work and advanced education. Teacher-librarians turn the age of information into the age of understanding, critical thinking, and learning to learn:

> For any of the learning specialists and, in particular, the teacher-librarian, the curriculum of the specialist is being integrated with the learning standards required by the classroom teacher. This "just in time" and "need to know" instruction helps learners build their knowledge base and at the same time helps them learn even more efficiently. Examples might include how to judge the differences between fact and opinion as a political issue is being explored; how to think critically about conflicting media messages encountered on the topic; how to paraphrase by selecting major ideas in a variety of texts; and how to use a wiki to collaboratively build a case for a position the group is creating. As these learning journeys happen,

"In order to produce successful learners in the 21st century, schools must integrate information literacy throughout the whole curriculum."

"For schools, the challenge is not only to create the networks, acquire the software, and make both operational, but to react to the transformative influences of technology on the way everyone learns."

the adults are watching, coaching, and assessing progress to insure that every learner either meets or exceeds the learning expectations. In other words, the Learning Commons supports a school-wide culture of inquiry fostering "habits of mind" and "learning dispositions" conducive to success. (Loertscher, Koechlin, & Zwaan, 2008)

COLLABORATION KEYSTONE

The establishment of the Learning Commons as a community of learners opens the door for more effective instruction and, consequently, school improvement. Here we experience many types and layers of collaboration—everyone working together to analyze and improve teaching and learning for all. Teachers and administrators work on specific facets of school improvement and safety. Students work with other students and teachers on solving problems, building knowledge, and creating together. The broader school community works within the Learning Commons to support learning and local initiatives. All work together supported by the rich resources and technologies of the Commons. (Loertscher, Koechlin, & Zwaan, 2008)

The learning commons serves as the center of a collegial schoolwide effort to improve teaching and learning. It is a learning laboratory where new teaching and learning strategies are developed and taught; where experimentation with new strategies is tested and analyzed; where action research is used to verify instructional progress; and where successful experimentation then radiates out into the rest of the school and is showcased to boards, parents, and the community. The leadership team for this effort includes administrators, selected classroom teachers, specialists in the school (music, art, PE, etc.), specialty coaches, librarians, and technologists. This leadership team schedules the learning commons with the major "experiments" and experiences throughout the school year so that a constant stream of learning initiatives are beginning, being implemented, assessed, reported, and applied throughout the whole school. The result is school-wide improvement through the learning commons.

TECHNOLOGY KEYSTONE

The Learning Commons is the space where learners and technology converge. This merger creates a dynamic environment where world class learners blossom. There is general agreement that learners who are astute in the wise use of technology have a better chance of competing globally. For schools, the challenge is not only to create the networks, acquire the software, and make both operational, but to react to the transformative influences of technology on the way everyone learns. (Loertscher, Koechlin, & Zwaan, 2008)

Students know how to operate the various technologies in the learning commons to their advantage, and they also know how to maximize their own efficiency and collaborative learning using that technology. They apply their technological expertise to design innovative products and compelling presentations to communicate

The learning commons serves as the center of a collegial schoolwide effort to improve teaching and learning. It is a learning laboratory where new teaching and learning strategies are developed and taught; where experimentation with new strategies is tested and analyzed; where action research is used to verify instructional progress; and where successful experimentation then radiates out into the rest of the school and is showcased to boards, parents, and the community.

FIGURE 3

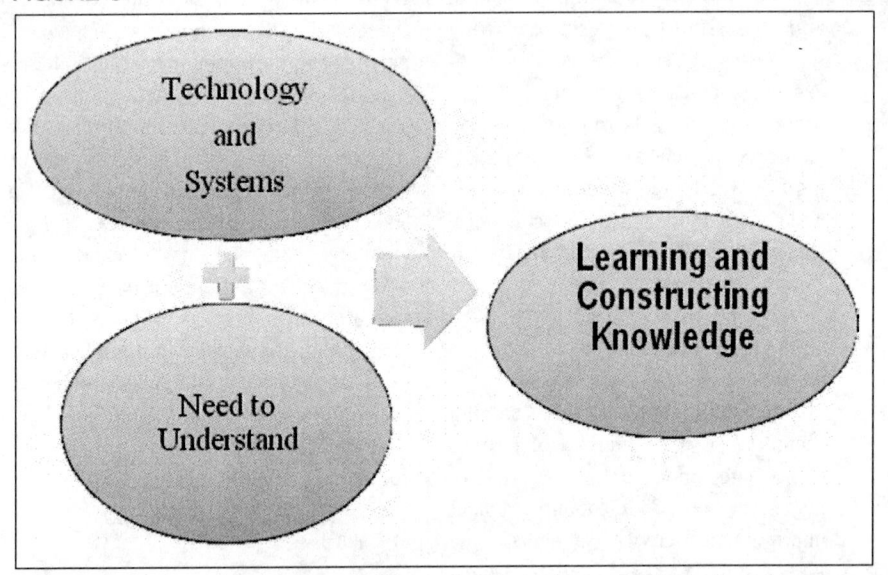

- Technology and Systems
- Need to Understand
→ Learning and Constructing Knowledge

For teachers, the library as a professional development center is the keystone of learning to teach effectively with technology as the technological environment of the school develops and innovation occurs. To ensure real-world learning experiences, the learning commons prepares for change as technologies and needs shift.

their new learning. Learning is maximized and deepened as the technology becomes transparent.

In an age when young people's social networking skills often surpass those of adults, teacher-librarians utilize these known skills and transfer them to academic situations. Children and teens are introduced to the concept of building their own information, organizational, and group spaces using Web 2.0 technologies. They not only learn to build and manage these information spaces but also learn how to manage themselves in those spaces (Williams & Loertscher, 2008). This emphasis on client-centered technology allows young people to participate in building smaller, high-quality, useful, and safe information environments.

A LEARNING COMMONS MOMENT—ASK THE LEARNERS

If a visitor to the school were to randomly select a table during lunch and ask students about the role the learning commons plays in their school life, how would students respond? What sort of information would the students share? Perhaps something along these lines:

- Environment—a comfortable place where they can work, relax, learn, create, or do.
- Access—a convenient, 24/7 source of materials, information, and advice they trust.
- Assistance—a place to comfortably obtain help from both adults and fellow students.
- Personal Contributions—a place to voice opinions and give advice to assist in decisions about construction of the learning commons; a place to make contributions and feel some sense of ownership.
- Experimentation—a place to try new things, test technology or software, develop special projects, and see the adults doing the same.
- Technology—a place to access and use hot new technologies and programs/software; a source of connection to the digital world and a center for discussion about that world and how they control it to their advantage.
- Activities and Exhibitions—a variety of activities they have participated in or seen happening; many student productions that are part of the digital museum of the school.

Certainly they would convey a sense that adults coach and mentor them when they need help and that staff inquire about how they learn as well as what they know.

They would talk about a caring, supportive place to learn without angst and pressure.

In other words, the various learners recognize that the learning commons is a client-side organization where they have some say in what goes on and they are contributing as well as receiving as a user. They may not understand the impact that the learning commons is having on teaching and learning throughout the school, but they should recognize they are engaged as they inquire, use, contribute, work, and create.

A LEARNING COMMONS MOMENT—ASK THE TEACHERS

Likewise, if a visitor to the school were to enter the teachers' lounge and interview random teachers about the learning commons, what sense of its value would be expressed? Perhaps something along these lines:

- Environment—a part of their classroom, an extension of both work and learning activities.
- Access—the 24/7 source of materials, information, and advice they trust; a place to send individuals or small groups or schedule the entire class there as needed.
- Assistance—a place where they obtain help from both adults and students who are sharing their expertise.
- Personal Contribution—a place to voice opinions and give advice to assist in decisions about construction of the learning commons; a place to make contributions and feel a sense of ownership.
- Experimentation—a place to learn, test, and share new strategies, test technology or software, and develop special projects; the center of professional development.
- Technology—the recognition that the learning commons is the source of their connection to the digital world that extends into their classrooms.
- Activities and Exhibitions—a place for a variety of activities they have seen

> "They [students] may not understand the impact that the learning commons is having on teaching and learning throughout the school, but they should recognize they are engaged as they inquire, use, contribute, work, and create."

> "(In other words) teachers recognize the advantages of building and maintaining a client-side learning commons and feel at ease in the give and take of the idea of the experimental learning center."

happening and knowledge that their students' work and productions are a part of the digital museum of the school.

- a place where they do not feel they are alone in the challenge of elevating every learner toward excellence; a place to be part of a teaching and learning team that merges classroom teachers and specialists in a mutual quest.

In other words, teachers recognize the advantages of building and maintaining a client-side learning commons and feel at ease in the give and take of the idea of the experimental learning center.

The learning commons is long overdue. Begin the work today. Reinvent your school library and computer labs; listen to your clients; build learning partnership teams; infuse the best teaching science. Be prepared for dramatic results!

AN INVITATION

We need to hear your stories as we all join forces to reinvent school libraries and computer labs. Please participate in our wiki at schoollearningcommons.pbwiki.com.

REFERENCES

Loertscher, D. V., Koechlin, C., and Zwaan, S. (2008). *The new learning commons: Where learners win! Reinventing the school library and the computer lab.* Salt Lake City, UT: Hi Willow Research.

Williams, R., and D. V. Loertscher. (2008). *In command! Kids and teens build and manage their own information spaces.* Salt Lake City, UT: Hi Willow Research.

Carol Koechlin and Sandi Zwaan have worked as classroom teachers, teacher-librarians, educational consultants, staff development leaders, and instructors for Educational Librarianship courses for York University and University of Toronto. In their quest to provide teachers with strategies to make learning opportunities more meaningful, more reflective, and more successful, they have led staff development sessions for teachers in both Canada and the United States. They continue to contribute to the field of information literacy and school librarianship by coauthoring a number of books and articles for professional journals. Their work has been recognized both nationally and internationally and translated into French, German, Italian, and Chinese. They may be contacted at *koechlin@sympatico.ca* and *sandi.zwaan@sympatico.ca*.

David V. Loertscher is coeditor of *Teacher Librarian*, author, international consultant, and professor at the School of Library and Information Science, San Jose, CA. He is also president of Hi Willow Research and Publishing and a past president of the American Association of School Librarians. He may be reached at *davidlibrarian@gmail.com*.

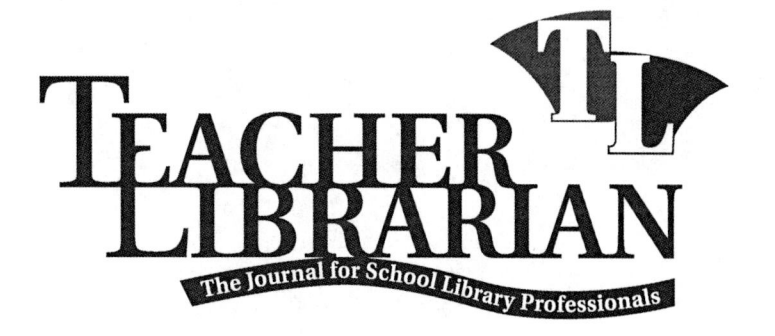

Part II:

Learning Commons Examples

One of the first questions about the Learning Commons is that while the theory is interesting, where could ~~one find~~ an actual Learning Commons be visited and seen in actual operation.

Admittedly, a major change in foundational ideas in education is difficult to implement at the outset. However, unbeknown to the editors, Valerie Diggs in Chelsford, MA was in the process of developing her high school library as a Learning Commons. In her article, she describes her thinking as it evolved over several years from a change in program and then followed by the transformation of the physical environment.

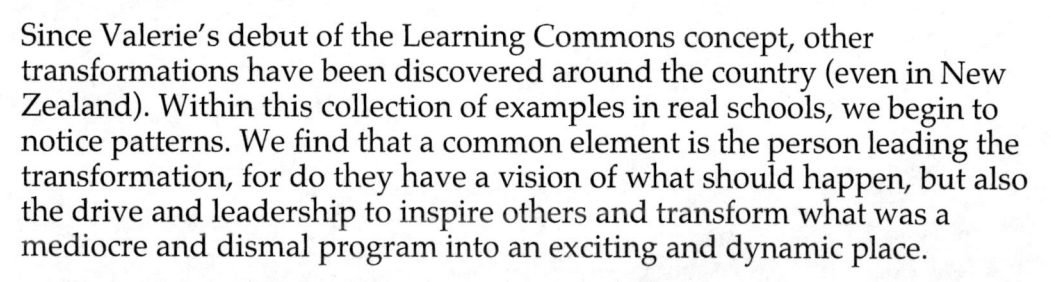

Since Valerie's debut of the Learning Commons concept, other transformations have been discovered around the country (even in New Zealand). Within this collection of examples in real schools, we begin to notice patterns. We find that a common element is the person leading the transformation, for do they have a vision of what should happen, but also the drive and leadership to inspire others and transform what was a mediocre and dismal program into an exciting and dynamic place.

Our own visits to such places convinces us that the major transition is when ownership of the Learning Commons transfers from the teacher librarian or the teacher technologists over to the faculty and the students. Such a transfer of ownership is probably the most important characteristic of those places you are invited to read about.

Readers are also encouraged to go to the wiki at http://schoollearningcommons.pbworks.com and find there a number of presentations made by practitioners. Many articles and books that encourage transformation can be found referred to there, and there is a place to share your own experiences as you transform your own school.

FEATURE ARTICLE

from library to learning commons: a metamorphosis

In August 2008, I published with Carol Koechlin and Sandi Zwaan published a book proposing to the school library community a major shift in the foundational ideas of the school library. We proposed a shift to a learning commons concept based on client-side principles. Unknown to us, Valerie Diggs had been doing such a transformation in Chelmsford, MA, and we visited her learning commons for its dedication in December, 2008. We asked Valerie to document the development of the learning commons in her school as a case study for major change in the concept of the school library. The following is Valerie's account but editorial comments have been placed throughout to provide some analysis of how teacher-librarians can actually move into the center of teaching and learning. Here is her account:

RECENTLY, THE CHELMSFORD HIGH SCHOOL LIBRARY (CHSL) IN MASSACHUSETTS UNDERWENT A TRANSFORMATION THAT WAS NOT ONLY LONG OVERDUE IT WAS ALSO A METAMORPHOSIS THAT WAS TO HAVE TREMENDOUS EFFECT ON THE STUDENTS, ADMINISTRATORS, TEACHERS, AND COMMUNITY MEMBERS.

While I knew the physical transformation of our facility was important and would certainly have a positive effect on the school, I did not expect the response it received from the community. We unveiled the new space in a special celebration that drew the press and such giants in the field as Dr. Ross Todd, an associate professor of library and information science at Rutgers University, and Dr. David Loertscher, professor at San Jose State University, who came to see what we had done.

The *Oxford English Dictionary* defines metamorphosis as "the action or process of changing in form, shape, or substance; *esp.* transformation by supernatural means"). The slow metamorphosis of CHSL into a space we now call a Learning Commons was deliberate and involved substance, form, and shape; and as for the supernatural—well we think so!!

This is a story of transformation in the truest sense, from the traditional media center to the not-so-traditional Learning Commons now occupying the third floor of Chelmsford High School, right where the library used to be.

I would like to be honest and open; I did not arrive in school one day and say: I think I would like to renovate this physically tired facility known as the library and, while doing so, I think we will start calling the space a "Learning Commons". There is much more to this transformation than just a name change, new paint, carpeting, and furnishings.

Perhaps I can begin to explain what happened by using this brief description of metamorphosis: "What triggers metamorphosis is largely unknown, although environmental factors are often involved" (Burkett 2009). There were numerous environmental factors involved in the decision to renovate the library. I must make it clear, however, that the condition of the physical facility was subordinate to the programmatic changes that made this transformation one of substance and meaning.

What, then, exactly did happen?

We must recognize that program is paramount in the foundational principle upon which Valerie sets off on her long journey toward the center of teaching and learning. We will draw attention to significant markers along the journey. As you read this case study, make your own list of important leadership principles and then compare your list with the ones we notice in our comments.

BACKGROUND

Chelmsford High School was built in 1971. Then principal, George Simonian, whose tenure at CHS was nineteen years long, presided over the planning and construction of the high school with a

by valerie diggs with editorial comments by david v. loertscher

FIGURE 1   FIGURE 2

Before the transformation

tenacious hand. He ensured, to the best of his ability, despite being limited by funding, that all programs had adequate space. One of the programs was library services. Back then, the library was called the *Instructional Media Center* (IMC), and 12,500 square-feet were set aside in the middle of the building which included a large workroom and office.

The IMC was state-of-the art for its time. The storage room housed 16mm Elmo projectors to show the numerous reels of film in the library's collection. Film loop players and film loops were stacked neatly on bright yellow metal shelving and filmstrip projectors of many types were also kept there, along with hundreds of boxes containing little round plastic filmstrip containers. Vinyl records lined the slanted yellow metal record bins framed by case after case of record players, all with neat black handles. As technology slowly changed, VHS tapes began to fill the empty shelves.

In 1971, the Building and Equipment Section (BES) statement of the function of a school library or IMC was the 1969 standards for school libraries published by AASL. These standards envisioned, like the 1960 standards had, a place where the full range of books and multimedia were made available to the school as a whole. At the time, women were generally hired to build library book collections and men were hired to handle the new array of audio-visual equipment and materials that were emerging rapidly. Thus, the principal was taking a bold step in merging all formats in a single location in the school.

In the library space, books were housed on the same bright yellow shelving with walls painted to match. Most of the shelving was 84" high and positioned to block the natural light from the library's rear windows. The six rooms on the periphery of the IMC, built to accommodate student group works, were barricaded with wrought iron gates and locked to prevent students from misbehaving within the glass-enclosed walls. The same wrought iron was used to fortify each of the two main entrances to the IMC.

While the vision of the merger was made between books and AV, note that the facility is constructed with tight control in mind and signaled a suspicion of teen behavior. Bright colors were thought to counter such anticipated attitudes. However, the facility's ambiance signals negativity.

TRANSFORMATION OF THE PROGRAM

When I was hired to be the high school librarian in 2002, I began to seriously look at what we were offering students as they entered the library. Before my arrival, study halls were eliminated with the introduction of block scheduling. Study halls had been held in the library because of lack of space elsewhere. However, those structured classes were eliminated and teachers were free to bring classes into the library at anytime.

Then the District of Chelmsford received a grant from Sun Microsystems for 60 desktop computers, 40 of which were placed in the high school library. The introduction of computers brought more teachers and students into the library.

Notice the very late introduction of computers into the library. For whatever reason, it is fifteen years behind in the adoption of a new information world. However, upon adoption of current technology, interest begins to revive for both students and faculty. It is now five years until the actual emergence of the new learning commons.

ALL ABOUT THE PROGRAM

However, the mere existence of computers and available space is not enough to meet the students' programmatic needs. On my arrival, I began to work more closely with teachers. They became familiar with my questions, requests for assignments as well as requests for time to look closely at assignments to determine how I could help. With the introduction of two new courses in the English Department, *Writing for High School* and *Writing for College,* as well as a deliberate movement by all departments to require more writing, I became more and more involved with the curriculum and instruction. Student learning became the focus behind everything we did, and teaching information literacy skills in the library, in computer labs, and in classrooms flourished. Our library began to play a key role in students' literary lives and was central to their learning experiences. Teachers began thinking about projects with the library in mind. The culture of teaching and learning changed over time to include the library as a major player. Was not this enough? Is not this what all school librarians strive for?

At this point, the facility with its negative ambience is 30 years old. The introduction of computers provides the excuse for Valerie to make a major push into the center of teaching and learning in spite of barriers both real and perceived.

What were we offering students and teachers? Reliable information? Yes. Technology? Yes. Help with assignments? Yes. I began to think this was all we could do. Yet something nagged at me. There had to be more; but what?

Note that Valerie is uncomfortable as she tries to turn stuff into productive learning. For the next five years she is constantly trying new ideas and she provides a list of things that begin to work. . Undoubtedly here were failures along the way, but persistence and creativity will pay off.

With one small step at a time, I started to build an assortment of events, ideas, and ways of doing things into our program. One of the first changes we made was to serve coffee one day per week in the library. Every Wednesday, from 6:45 a.m. to 7:24 a.m., coffee, hot chocolate, tea, and often breakfast-type foods were available. We developed a collaborative arrangement with a local coffee shop that allowed us to purchase their coffee and cups at a discounted price and the Chelmsford High School Library's *Java Room* was born.

On Wednesday mornings in the library, pots of steaming coffee and hot water for tea and hot chocolate were lined up on the aging classroom desks, which were placed side-by-side in the center of the library to serve as a reference desk. Trays of pastries and bagels donated by generous Chelmsford businesses beckoned the hungry students and staff members. The lines were long, the laughter loud, and soon the gathered students were enticed by the books on display to browse and check out books while they waited for their hot drinks. Students also sat and talked while teachers and administrators readied themselves for a day of teaching and instruction by recounting events of the previous days and catching up on daily news.

It's the old bait and switch technique. You come for coffee and end up with a book. But a large change is taking place. The culture of the place is turning around. The direction is toward a client-side focus.

Almost at the same time, the Chelmsford Public School community began a professional development initiative to introduce the concept of Professional Learning Communities (PLCs) into all seven of its schools. The theory behind PLCs is a perfect fit for any library program. Simply, a PLC encourages teachers to work together collaboratively. The book *Reinventing Project-Based Learning* tells us that "Creating a professional learning community means making time for new ways of working with colleagues" (Boss, Krauss, & Conery 2008). The concept of this *new way* had come to CHS; new to some, familiar to others, and frightening to a few. From the library's point of view, it was a welcomed initiative. What a perfect fit for what we had already been doing with teachers: collaboration with a focus on student learning and results (DuFour, 2008). I began to think of ways in which different departments might work together.

Notice that Valerie does not invent ways to move toward the center of teaching and learning. Rather, she links her vision to a major school improvement initiative. Her client-side initiative is not thought of as an add-on but a collaborative push across the school.

SOCIALIZE THE LIBRARY?

For example, the fine arts department consists of many talented teachers and students, but I had not had the opportunity to work with them other than on a very superficial level. Why not draw on the talents of this department and offer these students a venue to showcase these talents in a positive way? I met, talked, and brainstormed with some of the staff members of the fine arts department and our discussions led to the birth of the *Listening Lunches* program.

Named by the students in the fine arts department and supported by both the library and fine arts staff, this new collaborative program changed the culture of the high school and has become central to the life of the school. On at least one Friday per month, students pour into the library carrying trays of food, bottles of water or sports drinks, and paper bags bursting at the seams with sandwiches, fruit, and cookies. On Listening Lunches day, students love the opportunity to have lunch in the library while listening to their peers read poetry, sing songs, perform skits, play musical instruments, and more. Large rolling barrels stand ready to accept half-eaten lunches and empty bottles; round tables adorned with tablecloths and vases of artificial flowers are arranged in the back corner of the library; and rows of red folding chairs are usually stacked awaiting the throngs. For two and a half hours the library is filled with the sounds of student talent. Students and teachers sit, stand, eat, listen, and socialize in the space called the library.

Clever. Clever. The change in culture is not Valerie's program. It pushes the fine arts into the center of school culture. But what about the food, drinks, and noise? And all of this on the new carpet that is in Valerie's future?

Socialize in the library? Here is a concept often thought to be in comparative opposition to learning. Henry Jenkins, in his white paper entitled *Confronting the Challenges of Participatory Culture: Media Education for the 21st Century*, speaks of students in a participatory culture, one which requires a new set of literacies that "...almost all involve social skills developed through collaboration and networking" (Jenkins 2006). One of these skills Jenkins describes as "**Collective Intelligence**"—the ability to pool knowledge and compare notes with others toward a common goal" (Jenkins 2006). The school library should give students the space to work with each other, I thought. A place to go that offered technology, some privacy, and the atmosphere conducive to learning experiences. Someone nearby to answer questions could also help.

With an aging facility and rooms originally meant to be group study rooms now full of old AV equipment and locked to students, helping our students work in groups, collaborate on projects, and develop this skill of *collective intelligence* was difficult, at best.

The new image is not just culture or a social place. Valerie is worried about the library's central role in learning and she reaches out to the best theorists in education for ideas.

TRANSFORMATION OF THE FACILITY

As hard as I tried to convince our administrators that renovating the library would

16

APRIL 2009 | 35

benefit every student and staff member in the school, my detailed reports, requests, budget submissions, and pleadings were in vain. All I was hoping to do was transform a facility that was as dreary and tired.

While I was busy trying to convince everyone, the District of Chelmsford began a 31-million dollar renovation plan, which included plans for a new performing arts center, new science wing with sparkling new labs, technology classrooms, and instructional spaces. The two middle schools received new libraries: 5,000 square-foot wonders that soon became the "place to be" in both buildings. Other schools were outfitted with new flooring, roofs, boilers, etc. Through all those renovations and building projects, the high school library remained as is: a tired, 34-year old space, with duct-taped carpet, bright yellow shelving and walls, and desks with broken drawers and peeling facades. I admit, I did whine. But, whining did not seem to work. Now what?

How does Valerie inch up the priorities ladder for a facility to match the major change in her program? The squeaky wheel is beginning to turn.

The Town of Chelmsford hired a new town manager, Paul Cohen, who had been taken on a tour of the town's buildings and facilities by the search committee during the interview process. He saw the high school and its new science wing, the new performing arts center and the two beautiful new middle school libraries. The search committee did not dare bring him near the high school library. It had become so embarrassing a space that avoiding it was an accepted practice.

However, in May 2007, I invited Paul Cohen for a special visit to the library. To hear him tell the story, it was the first time he had ever been called by a school librarian in his over 17 years of public service. I gave him a tour, served him coffee in the workroom, and told him all about our programs, services, and student learning experiences. He saw for himself not only the duct tape but the entire, sorry mess. That fall, the capital budget was announced and included over $200,000 for the renovation of the high school library. In this case, a tour was worth many thousands of capital funding dollars. Our *Learning Commons* was on its way.

When the frontal assault does not produce results, Valerie does an end run. Notice that we are now five years into the transformation.

Many people have asked me how I was able to create such a space. How did I decide to call the library a *Learning Commons* and why? My response is that I did not decide; the program did. A recent article in *Teacher Librarian* describes the vision of a true *learning commons* as "...the showcase for high-quality teaching and learning—a place to develop and demonstrate exemplary educational practices. It will serve as the professional development center for the entire school—a place to learn, experiment with, assess, and then widely adopt improved instructional programs..." (Koechlin, Zwaan, Loertscher, 10). Our program defined us and the definition of what we did every day transformed our space into one that we now call a *Learning Commons*.

Notice that the vision for improvement keeps evolving and pointing toward excellence. It is a path to constant school improvement that is really never finished.

TEACHING AND LEARNING IN THE LEARNING COMMONS

A Spanish Honors Six class is investigating the current immigration policy in the United States. What do they need to know about our policies and why should they even want to know? These senior students came to the Learning Commons with their classroom teacher to find out.

At this point, Valerie could have taken us on a tour of architects, design, problems with vacating the library and construction, etc. Rather, we are taken on a tour of what this space is designed to accomplish—the tour of a fine learning experience. The facility must support the central focus of the program.

COLLABORATION WITH THIS NEW MODEL

They came into the Learning Commons interested and ready to ask questions. How did this come about? Why was I not facing a group of disinterested students with the typical "Can't wait to get through this assignment" look on their faces? It is because, prior to coming into the Learning Commons, the students spent time with their classroom teacher discussing their own roots as a way of making the lesson personal. That got them hooked and interested. Then they read a story called "Cajas de Carton," (Cardboard Boxes), a true story about illegal migrant workers in California in the 1950s and discussed it as a group.

The students then participated in an activity to consider four different positions the United States could take regarding the immigration problem, and were encouraged to come up with a fifth option. The learning and engagement with the topic continued when the students listened to three songs about immigration and saw the respective music videos. They discussed the songs and videos in small groups and voice recordings were made of the discussions for future reference.

In her own words, the classroom teacher Merrie McIvor said she got what she wanted from the activity; she "wanted to have them consider immigration seriously and individually" and "wanted them to write something" (personal email communication, January 7, 2009). So, the students were left to develop answers to their own questions. The activities leading up to the writing helped the students refine and define exactly what they needed to know and why they needed to know it. The short story, the songs, and their brainstorming of alternative positions taken by the United States on future immigration policy helped these senior Spanish students prepare. Not only did we have unique papers, but we had students were invested and interested in what they were asked to investigate. The resulting inquiry process was a natural progression.

FOR AN INFORMED AND STIMUTATED STUDENT

Organizing inquiry-based units around question development is a practice described by Wiggins and McTighe in *Understanding by Design* as being essential to providing "...teacher and students with a sharper focus and better direction for inquiry." They go on to tell us that developing personally meaningful questions "...render the unit design more coherent and make the students' role more appropriately intellectual" (2005, 27).

In our daily practice, how often do we see units based on essential questions, particularly those developed by students? How many units or projects such as the one on current-day immigration policy attempt to inform and stimulate students and bring them to the point where they can ask the question, and know why it is they need to ask it? There are too few such experiences for students today. What can we, as library media specialists, do?

The answer to this question is nothing new, and certainly nothing any library school student or current practitioner hasn't heard time and again. It is collaboration. Not simply pulling books, or book-marking web sites, or even creating pathfinders. But it is sitting down with teachers and saying "How can WE improve on this unit so our students can learn not only more, but better?" It is the "WE" in this equation that is important. Students will do only what is required of them, they will think thoughts only as deeply as we require of them. It is OUR job, alongside the classroom teacher, to offer our students today the opportunity to think critically and develop questions to answer that they really want to answer, questions that will lead them to turning information into knowledge and subsequently that knowledge into wisdom for a lifetime.

SIGNS DO MORE THAN POINT THE WAY

In our newly renovated Learning Commons, a quote from John F. Kennedy is drawn across the wall of the group work area: *We set sail on this new sea because there is knowledge to be gained.* It is with this mindset that we (and I include teaching staff, students, administrators, and library staff) conduct business. Knowledge can only be gained through the process of true and unadulterated inquiry. The inquiry process is crucial to our students' experiences and central to the culture of our space. According to Loertscher, Koechlin, and Zwaan, "Inquiry in the Learning Commons is a dynamic learner centered process" (2008).

Then, above our central information desk are the words "*Ask, Ask, Ask*" and in the Café area the words "*Think*" and "*Create*" appear above the counter-top seating. We encourage our teachers to make learning meaningful by requiring students to think, ask, and then create. Learning becomes meaningful and lasting, and students come away with a wonderful skill; the ability to think for themselves.

Signage has always been important in libraries, but in this one, we are not directed to things or equipment; we are encouraged to learn. The first thing you see when entering the learning commons sets that new ambience.

THE LEARNING COMMONS AS A CULTURAL CENTER FOR STUDENTS

Groups of students sit in a circle on our new eggplant-colored lounge chairs, discussing the beginnings of World War I and how they would like their project to look. Smaller groups of three or four students sit in the restaurant-style booths completing projects, working on homework, and collaborating on assignments. Some students sit on the cafe-height tables and talk among themselves, while others use the counter. Students are able to check out one of 29 laptops received as part of a grant from a large technology firm in our community. These laptops are wireless, and connect to the Internet anywhere in our *Learning Commons*. The group workrooms have been opened up, providing space for students to work together and for staff to meet.

A remedial reading class gathers each day to read a book of their choice, while sprawled comfortably on the soft furniture. Their teacher could easily have kept them at desks in the classroom, but she understands the students are much more at ease with themselves in the act of reading by sharing our space, the Learning Commons. Classes of foreign language students file through the Learning Commons to examine our art exhibits, take notes, and absorb the immensity of the works.

After school hours find the Learning Commons busy accommodating many students. Club members meet here, groups meet to complete projects or assignments, and peer tutors instruct at our tables, while other students find a quiet spot to read or think. Every day when we give the five-minute warning for closing, the announcement is always greeted with groans and pleadings of "just five more minutes." If the budget allowed, our closing time would be much later than the hour-and-a half we are now open after school.

One gets the sense that the learning commons is not only a flexible learning space but one that is full all day long with multiple groups and individuals using a very welcoming space that is both social and intellectual.

The fifty-eight inch flat-panel LCD TV mounted on the wall across from the main information desk serves to enhance our connection to students and staff by providing informational, entertaining, collaborative and educational messages. We use the television to showcase the projects and talent from all content. As they enter or leave the Learning Commons, students may check out our daily schedule, watch and listen to music videos created by the graphic arts students, see public service announcements produced by our health classes, view slide shows on just about anything, and absorb breaking news through a crawling banner. In our media-saturated environment, digital signage captures the attention of students. It is also the perfect medium to capture the wide variety of products our students create in the Learning Commons as well as throughout the school.

Capturing and archiving student creations reinforces the notion that if they help build the learning commons, they will use it. Ownership transfers to both teachers and students.

COMMUNITY SUPPORT

The new Learning Commons has been the recipient of continuous support from community members both within and without the school. Our grand opening event was held on December 5, 2008 with huge attendance, a long list of speakers, and a virtual landslide of donations of food, time, money, and wishes for success. From our principal, to principal emeritus, to town manager, chairperson of the board of selectmen, state senator, school committee chair, to library gurus from across the nation, as well as teachers, parents, community members, and fellow teacher-librarians from across the state, we were honored by everyone's presence. Our Learning Commons was launched, speeches were made, and the general consensus was that we had created a space that had become the center of learning. Our impact on the culture of Chelmsford High School, the teaching staff, and our students, was beginning to be felt by all.

Why will dignitaries, parents, students, and teachers flock to a learning commons? I was there. It was community pride, an electric sense of excellence, and opportunity to pay tribute both to a visionary teacher-librarian and to everyone who had participated in its creation.

The *Learning Commons* provides CHS students and staff members the opportunity to ask questions, think about answers, and create new meanings. We have become central to teaching and learning because our mission is tied to the mission and ideals of our school and district, and we are committed to offering our services and space to all of our constituents.

If we read any of the major works on leadership and innovation, we can find lists of characteristics that makes transformation possible. We have noted many of those characteristics in our marginal notes but there are others to think about and list. The end result in this case study is the turning from an organizational model to a client-side model. It is turning thinking 180 degrees from considering the needs of the user as subservient to organizational needs. And, if you ask Valerie if this learning commons has arrived, she will admit a certain sense of pride but with that nagging feeling that the evolution is still happening.

Our advice is to build your program first. This may take years to accomplish, as it did for us at CHS. However, remember a strong program is the foundation for a true Learning Commons.

For more information on the project, see the before-and-after slide show on the school's web site, http://www.chelmsford.k12.ma.us/chs/library/index.htm.

REFERENCES

American Association of School Librarians. AASL Standards for the 21st Century Learner. http://www.ala.org/ala/mgrps/divs/aasl/aaslproftools/learningstandards/standards.cfm (accessed January 7, 2009).

Boss, S., Krauss, J., & Conery, L. (2008). *Reinventing project-based learning: Your field guide to real-world projects in the digital age.* Washington, DC: International Society for Technology in Education.

DuFour, R., DuFour, R., & Eaker, R. (2008). *Revisiting professional learning communities at work.* Bloomington, IN: Solution Tree.

Jenkins, H. (2006). Confronting the challenges of participatory culture: Media education for the 21st century. Cambridge MA: John D. and Catherine T. MacArthur Foundation. Retrieved January 19, 2009, from http://digitallearning.macfound.org/atf/cf/%7B7E45C7E0-A3E0-4B89-AC9C-E807E1B0AE4E%7D/JENKINS_WHITE_PAPER.PDF.

Koechlin, C., Zwaan, S., & Loertscher, D. V. (2008). The time is now: Transform your school library into a learning commons. *Teacher Librarian. 36(1)*, 8-14.

Loertscher, D., Koechlin, C., Zwaan, S. (2008). *The New Learning Commons: Where Learners Win!* Salt Lake City, UT: Hi Willow Research and Publishing.

Wiggins, G. & McTighe, J. (1998). *Understanding by design.* Alexandria, VA: Association for Supervision and Curriculum Development.

Valerie Diggs is the Director of Libraries and Library Teacher at Chelmsford High School. She may be reached at *diggsv@chelmsford.k12.ma.us*.

FEATURE ARTICLE

Concord-Carlisle Transitions to a Learning Commons

Once you know a student, he or she no longer feels anonymous and is much more likely to ask for help.

ROBIN CICCHETTI

I n 2007 our library was dedicated to books, and it showed.

Concord-Carlisle Regional High School (CCHS) is in a high performing district 20 miles west of Boston with an enrollment of approximately 1400 students. Built in 1975, the CCHS Library is quite large with a three story open plan connected by ramps. It is the first space you see upon entering the main entrance of the school.

However, our library was a dark and cavernous book museum. Bookcases blocked so many of the internal and external windows of the library that whether one was trying to look in or out, the only view was of books.

The collection was quite large with 38,000 titles that had an average age of 31 years (we had books from 1976), which were spread non-sequentially over three floors. There was almost no way to find anything without a teacher-librarian taking you by the hand and leading you to the hiding place within the stacks. Microfiche machines sat cloaked under a thick layer of dust alongside thirty years worth of *The Reader's Guide to Periodical Literature* (2008), decades of magazine back issues, and obsolete card catalogs.

There was one cramped area for class instruction, which was not a problem since so few classes came in for instruction. Student use of the library was very traditional, focused on silent individual study. There were so many rules and punishments that the library was a place of continuous conflict. The only bright colors were from laminated signs that could be found everywhere in the room:

Please Do Not Move Chairs!
Penalty for Eating in the Library—After-School Detention!
Penalty for Drinking in the Library—Washing Library Tables!
Put Textbooks Back!
Silence!
No Group Work!
Don't Talk!

One had the feeling that the library would be a wonderful place if it were not for the students.

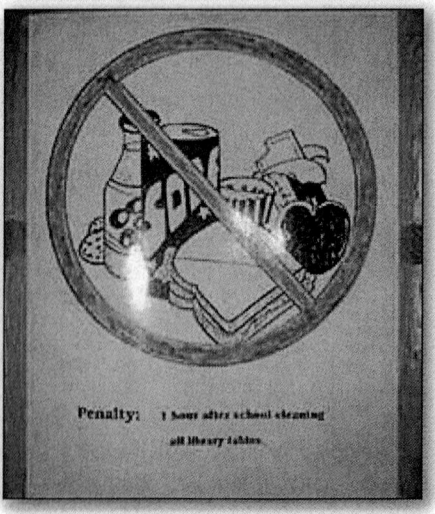

WE DECIDED TO CHANGE

In September 2007, I was hired as the CCHS teacher-librarian. From the very first day, I had the full support of an administration that thought as I did that the library could play a rather dynamic role as a learning commons—a center for learning and creativity for students and staff.

We are in the third year of our transition to a learning commons, a model that embraces the programmatic and space changes required to support learning information skills, critical thinking skills, collaborative work, creativity, and the joy of reading. Most important, it is a model where student learning comes first.

Here are the steps we have taken over the past three years to makes these changes, followed by a discussion of our results.

REINVENTING OURSELVES

Before we could even begin to address the physical needs of the facility, we needed to change ourselves. With the help of our very supportive human resources department we rewrote all job descriptions, focusing on student services and professional development. We went from having three generic assistant librarians to positions that covered specific areas of expertise:

- The description for my position, the school library media specialist, was adapted to reflect that as a teacher my content area is information and media literacy and I should possess the skills and technologies associated with those responsibilities.

- The reference librarian position became student services specialist, which meant reference work would now cover books, databases, web resources, online training, and tutorials.

- The position handling clerical tasks was changed to Accounts Specialist to include account administration and vendor communication, which was previously handled by the teacher-librarian. This made the accounts specialist position more critical and freed up time for the teacher-librarian to teach. Purchase orders and budget accounting also shifted to this job description.

- A new position of media production specialist was created to have a professional manage the partnership between CCHS library and our local cable television company. That individual keeps our technology department humming and provides advanced support for student digital and media projects.

Photo 3. The old main desk was not welcoming.

Our staff was totally re-energized as they bought into the changes. They are less disciplinarians and more professionals who actively engage in supporting student learning as an integral part of the school community.

FIX THE SPACE

While revising our job descriptions and responsibilities, we were also addressing the physical challenges presented by the space and the sprawling collection. The initial focus was on streamlining the collection and clearing the floor of extraneous junk. These two space goals would give us the ability to host classes more effectively, bring in light, and create extra space. Between the heaps of old furniture and equipment, the dust, and the signs everywhere, it felt like a cluttered prison.

Before moving the bookcases, we began an ambitious weeding campaign and removed over 5,000 books. We said goodbye to the microfiche machines, the *Reader's Guides*, the back issues of magazines, the card catalogs, and the excess 'stuff' that had been stored all over the library, and filled two twenty-foot containers for removal.

Our goal was to consolidate fiction on the second floor, with nonfiction and reference on the third, while placing everything into sequential order. This reorganization would allow students and staff to more independently navigate the collection and find material with guidance from, rather than reliance on, the teacher-librarian.

As we removed books, we reorganized the bookcases to clear space for students and classes. Without a budget for renovation, we moved some of the smaller bookcases ourselves. This impressed (and amused) our custodians so much they became great partners and moved the larger bookcases when they found time in their day.

Suddenly, the front entrance hallway of the school actually looked into the library and not just at 105.12–237.38. We could now see historic Walden Woods out of our second story windows and not Fiction A-G.

Tables were spread out, signs came down, and light that had been blocked streamed into a facility that suddenly looked very different.

GRANTS HELPED A LOT!

We wrote grant applications to local community groups and received funds for soft reading chairs, student supply carts, more than a dozen large potted plants, as well as new English and foreign language dictionaries.

Since then, our grant writing requests

Photos 4 & 5. Old Shakespeare has been replaced by urban art created by students

have expanded to include additional technology capabilities such as external hard drives for student work, higher quality digital video cameras, as well as light and sound kits for student productions. There are also plans for more new furniture as well as a moveable stage for student performances. Through the partnership with our local cable television company we have received training and support for the new media tools, and the result is an increased flow of student and school related media for broadcast.

NEW TECHNOLOGY

Our district had already invested heavily in technology, but little of it had ever made its way to the library. In 2007, the library had eight aging PCs. During my first year, we added ten desktop iMacs, a laptop cart of twenty Macbooks, and quickly installed a Promethean ActivBoard in the first floor instructional area. Each subsequent year has seen the addition of more technology and wireless capability.

We mounted a large LCD screen above the circulation desk and hooked it up to a computer to run a digital display. There was always something new scrolling to catch student and faculty interest such as school news, teacher news, student activities, new books, and library contest promotions. This was a powerful symbol that there were new things going on in the library.

Our staff development was critical during the first year as we added new computers and began introducing web-based tools to the school. Moving from PC to Mac was a big step and we spent time learning about the laptop carts and basic trouble shooting. The library staff was encouraged to take a laptop home for the weekend and 'play' with it, uploading photos and making movies with iMovie.

We reallocated space to create an eight-station, state of the art media lab. Students now work on media rich projects alone or in groups, for classroom projects in all disciplines.

Our filters opened as well. After initial debates on the merits and drawbacks of social networks, we gained support from our forward-thinking administration to allow access to Facebook and other platforms. This openness has not proved problematic and instead makes the library pertinent to our students' daily lives. They have a space where they can continue their busy online lives as they access their academic work supported by librarians who guide (and often troubleshoot) rather than police and constrain.

THE COLLECTION

Though book-buying budgets have been tight over the past three years, we have been able to add to and diversify our collection, particularly with the addition of graphic novels, urban literature, MP3 Playaways, Flip digital video cameras, and eBooks.

This diversification has been of particular importance in servicing the needs of our special education students. Alternate sources for information and texts including audio and graphic novel adaptations for visual learners, are helping tutors and students access academic content. Adaptive technologies like Kurzweil and other text-to-speech features are part of our growing suite of student services.

OUTREACH—FACULTY AND PROFESSIONAL DEVELOPMENT

Never underestimate the power of the cookie.

During that first year, I sent out a monthly group email to the various department chairs featuring a cookie of the month. The first department chair to invite me to one of their regularly scheduled department meetings got a double batch of cookies for their staff while I talked to them about the new library program. There was a bidding war the month I offered Whoopie Pies.

Administrators made sure I was told they felt left out and so we extended invitations for them to join the department visits. It was fun and an incredibly effective way to get in the door to promote the library, explore collaborative opportunities, and investigate databases specific to their content area.

The library became the site for professional development and now supports faculty in developing wikis for their classes and embedding Google Custom Search engines to guide students toward the best and most appropriate web-based resources for their curriculum content area. The professional development offerings we've provided through the library over the last two years has included instruction on Vokis, Web 2.0 Smack Downs, widgets, and specialized databases.

Most exciting is the rise in requests from faculty for assistance in advanced web searching and databases for class material and course content. This increase is evidence that the learning commons philosophy is influencing pedagogy. An additional benefit of providing professional development for faculty is that it creates opportunities for collaborative planning. Using the learning commons as the location for professional development gets teachers out of their classrooms, their departments, and their regular patterns. It is a neutral space dedicated to new ideas, new skills, and becoming a location for a new kind of collaborative learning experience.

COMMUNITY

Central to everything and more important than I realized back in 2007, is what a pivotal place a true learning commons becomes in the school community. Creating an environment of caring, genuine student support, and trust takes stepping outside your comfort zone on occasion, taking risks frequently, and trying new things always.

I also cannot emphasize strongly enough how important it has been for me to learn student names. Once you know a student, he or she no longer feels anonymous and is much more likely to ask for help. It also allows for an ongoing dialogue about how projects are going and a greater appreciation of student needs, challenges, and frustrations. It even allows me to engage in the endless Edward vs. Jacob debates (Twilight, 2008), adjudicate fascinating disputes, and laugh with the students as part of a community.

When we are asked by desperate students to chaperone a dance, we always try to accommodate. The students deeply appreciate when we chaperone their dances because it creates a common ground. This year, we are moving beyond dances and I will be chaperoning a trip to Japan. Additionally, the library's media production specialist will be traveling with a school group to Turkmenistan, to help document the trip on video. Traveling and learning with our students and other faculty members, as well as incorporating social media to share our experiences with parents will model the communication skills and behavior we want for our students.

Another wonderful partnership came about when a representative from our student senate approached us for help with school-wide elections. They wanted to move away from inefficient paper ballots and use an online tool. We happily checked out laptops and the elections occurred in the cafeteria during lunch blocks, monitored by students, with library staff rotating freshly charged laptops as needed. This initial collaboration has spawned many more opportunities for involvement in student life. As engaged partners, we can also guide students in collaborative, technical, and communication skills. Our integration into their community helps students feel that the learning commons is theirs.

OVERHAULING THE ACADEMIC PROGRAM

Most important of all was of course the transformation of the academic program.

Using standards from the American Association of School Librarians (AASL, 2009), our Massachusetts State and School Library Associations, as well as guidance from curriculum areas such as Framework for 21st Century Curriculum and Assessment from the National Council of Teachers of English (NCTE, 1996) and the National Council of Social Studies (NCSS, 1994), it was crystal clear we needed to improve our information and media literacy skills. While we overhauled the space, we also overhauled the program.

In 2007, there was little history of collaborative work. Teachers sent classes to the library when they were out sick, or they occasionally sent students down with the vague mission to "find a book." There were established research projects that required a visit to the library for a review of databases built into the lesson map, but there were no discussions about the goals of the class.

Teacher by teacher and visit after visit, we began to change the dialogue. We analyzed assignments for the skills we wanted students to build and tried to determine how these could be assessed. We suggested new tools for search, organization, and synthesis to incorporate more technology and provide differentiated instruction.

TECHNOLOGY AND OUR PROGRAM

Many teachers were nervous and afraid they could not possibly allow their students to use applications they did not know themselves. Partnering with teachers and providing close support has been important; many teachers are now trying new things with their students, confident we will support them.

Today classes that visit the library learn about our databases and deep web searches, but they also learn about tools such as Google Wonder Wheel. We teach numerous Web 2.0 tools as needed, such as widgets, Animoto, VoiceThread, Glogster, Prezi, Flickr Creative Commons, RSS feeds, iGoogle, PageFlakes, and other tools that will help students create and personalize the Web to their individual information needs.

For management and organization of academic projects, we teach platforms such as NoodleTools and Diigo. Online journaling with blogs and discussion forums using wikis are also taught and supported. Copyright, attribution, and creative commons licensing are frequent lesson topics as classes cycle back to the end of a research project to begin focusing on the synthesis of their work. We also teach students to discriminate among online sources and demonstrate critical evaluation of any source they use in their work.

Analyzing our database statistics and surveying the faculty led us to make some changes in our databases. JSTOR was added for teachers and students looking for higher level academic content, and we added access to primary source portals such as the National Archives. Oxford English Dictionary Online was added at the request of English teachers who wanted to delve deeper into language and etymology with students. Teen Health and Wellness was added to support life skills research projects for our junior Health and Fitness classes. The Concord Free Public Library has been a wonderful partner and hosts workshops in their Science Resource Center and collaborates to offer training for our Science Department and their reference staff.

The increased activity is not only about technology. Our circulation statistics are up in large part because requests for book talks have also increased. More teachers are bringing their classes to get ideas for independent reading projects. There is nothing as rewarding as a rush for the table after a presentation of the top, new young adult novels. Listening to teenagers talk about books and seeing them walk out with one in hand is as deeply satisfying now as it has always been for teacher-librarians.

THE RESULTS ARE FANTASTIC!

We have tracked, quantified, and documented the results of our changes. Using Goggle Docs we created forms to enter and aggregate data in a simple way. One quick way to assess our progress in building a space and program valued by the community is by tracking patron visits. Figure 1 shows the remarkable increase in the number of visits.

Circulation statistics for the same period also tell a rewarding story. Students are checking out books in substantially greater numbers each year.

Keeping tallies of visits by department tells us where outreach has been effective and to which departments we need to focus our support and energies. As of October 2009, we have excellent integration with the English department. Visits from our foreign language classes are higher than they were last year and social studies is only a bit lower. This is good information that will guide our plan for outreach throughout the year.

Tracking the technology skills we teach over the course of the year helps guide instruction and self-assesses our progress in collaborative planning with teachers. This

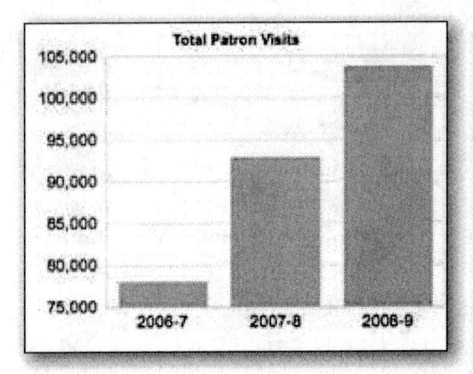

Figure 1

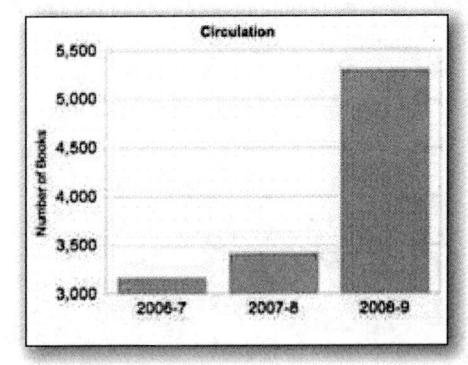

Figure 2

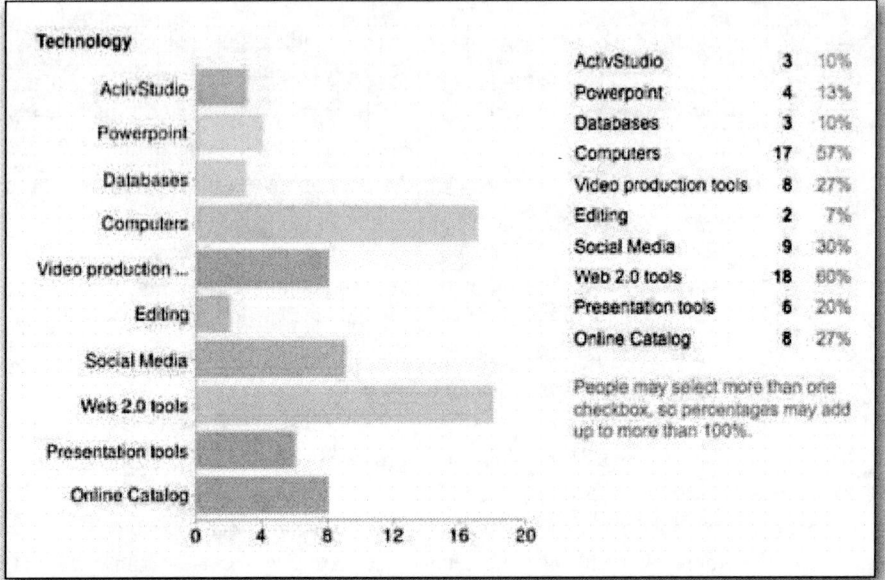

Figure 4

data reflects the integration of technology in class visits for September to October 2009. Students are using class time to access web-based resources for video production and social media platforms. Once research projects begin in January, we expect to see the database numbers increase. Data collection allows us to track, monitor, and respond as needed.

Finally, every class is matched to the AASL Standards (2007) and providing data on which skills are being addressed is very helpful. Student learning is the priority and documentation is evidence of the specialized instruction and support provided through our program.

WHERE ARE WE NOW?

Two and a half years into the process, our physical space now serves the many needs of our students, staff, and community. We have gone from being a warehouse of books to a busy, vibrant place for students and faculty to gather and learn in a connected and collaborative environment. The cost of this transition to our district has so far been minimal. It was done with sweat equity, small grants, and cookies.

Light pours in through windows onto students curled into comfy chairs with a laptop or playing Scrabble in groups at a table. Our wireless network is open and many students work on their personal netbooks or read on their iPods. There is ideal space for diverse creative learning, group work, and the various kinds of academic or personal creation and production. Double and often triple class bookings keep the entire staff busy, and student visits continue to climb.

We host Skype events—most recently with a humanitarian aid worker in Pakistan. Student art has replaced posters of authors on our walls and we have had art installations from our Japanese sister school in Nanae (http://www.town.nanae.hokkaido.jp/english/default.htm), as well as photographic exhibits from UNICEF. Our digital displays promote school activities and new clubs and groups regularly use the facility. After school, the Taiko Drum Club meets twice per week in our space, with the boom of the drums heard all over campus. Nothing symbolizes the change from a traditional library to a dynamic learning commons like Taiko drumming!

The CCHS Library has been transformed both in space and program from where it was three years ago. Yet, the changes will never stop occurring as we continue to adapt to the changing environment and the

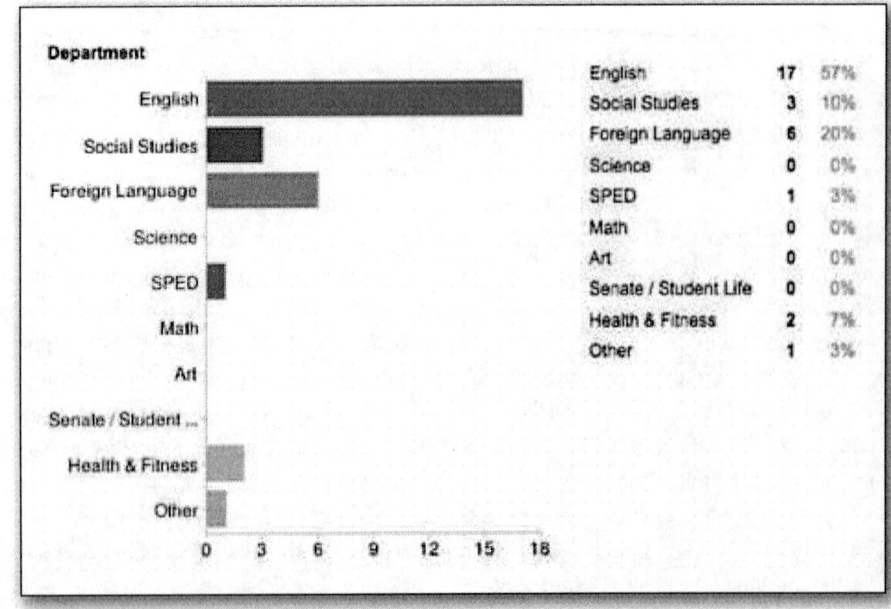

Figure 3

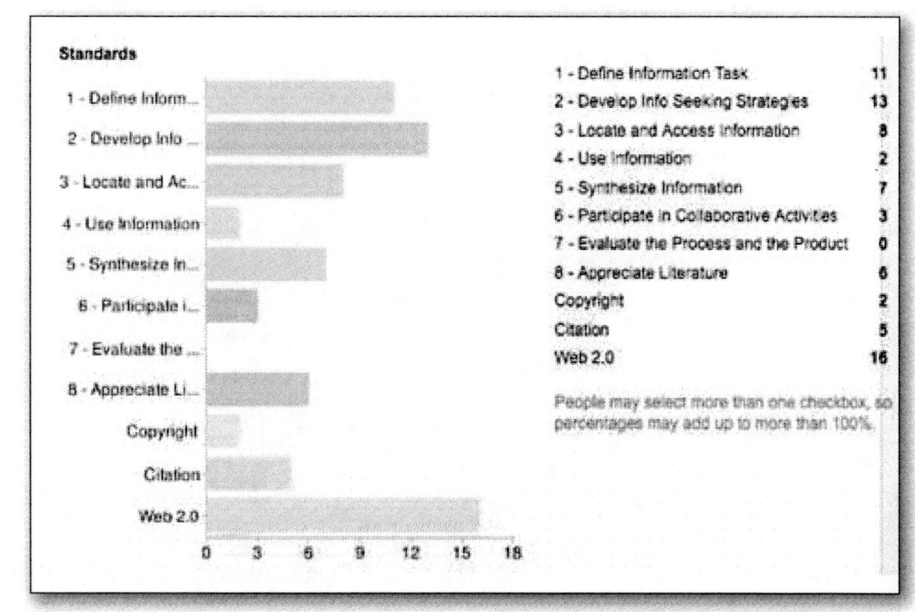

Figure 5

needs of our students. Next year though, we plan to mark our transition and officially rededicate our space as a Library Learning Commons.

Since attribution is a skill all librarians should model, I'd like to credit the following people for helping to provide the vision for our transformation. Special thanks to Valerie Diggs and her Chelmsford High School Learning Commons for blazing the trail, the Massachusetts School Library Association, David Loertscher, Joyce Valenza, Doug Johnson, and the many other teacher-librarians and educational technology bloggers who, via my RSS feed, provide me with the inspiration and professional development to bring the best to our students, every single day.

RESOURCES FOR FURTHER READING

American Association of School Librarians. (2007). *Standards for the 21st Century Learner*. Chicago, IL: American Library Association.

Burge, K. (2008). "New 'Learning Commons' defies commonplace." *The Boston Globe*. Retrieved November 15, 2009 from http://www.boston.com/news/local/massachu-

Photo 7: Look at the new Learning Commons now.

"Creating an environment of caring, genuine student support, and trust takes stepping outside your comfort zone on occasion, taking risks frequently, and trying new things always."

"Our integration into their community helps students feel that the learning commons is theirs."

"We have gone from being a warehouse of books to a busy, vibrant place for students and faculty to gather and learn in a connected and collaborative environment."

setts/articles/2008/12/08/new_learning_commons_defies_commonplace/.

Kenney, B. (2006). "Rutgers' Ross Todd's quest to renew school libraries." *School Library Journal.* Available from http://www.schoollibraryjournal.com/article/CA6320013.html.

Loertscher, D. V. (2008). "Flip this library: School libraries need a revolution." *School Library Journal.* Retrieved October 20, 2009 from http://www.schoollibraryjournal.com/article/CA6610496.html.

National Council for Social Studies Task Force. (1994). *Expectations of excellence: Curriculum standards for social studies.* Waldorf, MD: NCSS Publications.

National Council of Teachers of English and International Reading Association. (1996). Standards for the English Language Arts. Retrieved October 20, 2009 from http://www.ncte.org/library/NCTEFiles/Resources/Books/Sample/StandardsDoc.pdf.

Ontario Library Association. (n.d.). "Evidence based practice links." *Thank you can't? Yes you can! The teacher librarian's toolkit for evidence-based practice.* Retrieved October 20, 2009 from http://www.accessola.com/osla/toolkit/Home/EBPLinks.html.

Todd, R. (2006). "Guided inquiry: A framework for learning through the school library." Presented at East Asia Regional Council of Overseas Schools (EARCOS) Teachers' Conference, Edsa Shangri-La, Manila. Retrieved October 20, 2009 from http://cissl.scils.rutgers.edu/.

Todd, R. (2008). "The Evidence-Based Manifesto for School Librarians." *School Library Journal* 54 (4), 38-43. Available at: http://www.schoollibraryjournal.com/article/CA6545434.html.

Todd, R., & Kuhlthau, C. (2004). *Student learning through Ohio school libraries: The Ohio research study.* Columbus, Ohio: Ohio Educational Library Media Association. Retrieved October 20, 2009 from http://www.oelma.org/OhioResearchStudy.htm.

Trilling, B., & Fadel, C. (2009). *21st Century Skills: Learning for life in our times.* San Francisco: John Wiley & Sons.

Valenza, J. (2007). "Live from Treasure Mountain: My Rant." *Neverendingsearch. School Library Journal.* Retrieved October 20, 2009 from http://www.schoollibraryjournal.com/blog/1340000334/post/1200016320.html.

Valenza, J., & Johnson, D. K. (2009). "Things That Keep Us Up at Night." *School Library Journal.*

Wagner, T. (2008). *The global achievement gap: Why even our best schools don't teach the new survival skills our children need—and what we can do about it.* New York: Basic Books.

Robin Cicchetti is now in her 11th year as a teacher-librarian in Massachusetts and currently serves at the Concord-Carlisle Regional High School. She previously worked in school libraries in England and Switzerland. She may be contacted at *rcicchetti@colonial.net.* For the Concord-Carlisle Library Learning Commons Blog visit http://concordcarlislelibrary.blogspot.com/.

The Learning Commons is Alive in New Zealand

FEATURE ARTICLE

Our Library is sooo much more than simply a place to store a collection of books in the hope that someone takes them home to read them!

PEGGY STEDMAN AND GREG CARROLL

The Allen Centre, the Learning Commons of the Outram School on the Taieri Plains of New Zealand, was designed to provide stimulating support for the school's learning and teaching program.

Our Library is sooo much more than simply a place to store a collection of books in the hope that someone takes them home to read them!

The Allen Centre's mission is at the center of the school's Vision (2003):

"With our help, students must build strong foundations for their personal square. Inside this square, we guide them as they develop and organise a host of skills. But our most special challenge is to create an environment in which students grow powerful enough to burst outside the square. If we are successful, they will enter a galaxy of lively thinking, dreams and visions."

The vision for learning is well-understood by the staff, board of trustees, students, and community. This vision is depicted graphically in posters around the school, with the intention to "trigger flights of imagination and passionate explosions of ideas and interests" so students are motivated and empowered.

We have discovered, it is quite a different challenge to walk the talk. This walking is what we have worked very hard to do.

THE VISION THAT DRIVES US

The Vision continues to be a fundamentally important guideline for every single decision we make at Outram School. It enters conversations on a daily basis—a metaphorical yardstick that enters all debate. Its rather emotive articulation has been deliberate. It paints a mental picture, free from edu-speak conventions (which can become quite stale), and is intuitively accessible to the wider school community. Its strong visual interpretation resonates with even the youngest students who have no difficulty interpreting its meaning, and can verbalize it in a personally meaningful way.

Figure 1. This poster depicts The Allen Centre mission.

To translate this Vision into a real working space, we built a space where the book collection surrounds the walls, giving us an opportunity to fill the central space with all kinds of displays and workspaces for individuals and groups. It is a project space; a learning space; a busy space where the teacher-librarian connects with the classroom but also pushes beyond it.

By complementing the efforts of classroom teachers with the work of the Allen Centre, the learning commons staff fulfills the Vision by aiming to "generate a wealth of lively and accessible learning contexts... to acknowledge the notion of different learning styles and capitalize on the natural curiosity of children". Part of the work includes science and technology activities based in the hall and arts-workshop complex.

The displays and workshops featured on the Allen Centre Wikispace, http://allen-centre.wikispaces.com, define our 'point of difference', reflect this Vision, and serve several specific functions:

• They aim to excite and engage children and develop in each child a highly motivated personal learning culture, which over time becomes their norm.

• They aim to trigger and feed interest through a rich variety of learning contexts across all learning styles, capitalizing on the natural curiosity of children.

• They enrich classroom programs through tangential connections.

• They broaden the perception of a library from a quiet book collection to a vibrant interesting borderless space in which books are completely relevant and information "happens."

Through the Wikispace the walls of the Allen Centre become 'transparent'—you do not have to be physically present in the Centre to take part in the programs or see what is happening. It also enables 'time-shift' where children and families can interact with Centre activities even after the time they actually happened.

While seemingly serendipitous for the children, the displays and workshops are far from being ad-hoc in their design. They are very carefully thought through. The special challenge is how to present the ideas visually and interactively so they can be customized for and accessed by all age groups in a short amount of time at the start of each session, but also be stand-alone-engaging when the librarians are not around. Workshops are linked to the displays. Typically they run in blocks of several lunchtimes per week over several weeks. They are voluntary and open to every student regardless of age and it is not unusual to have 5-7 year olds working alongside 12-13 year olds on similar topics! Volunteers from the community are often in the Allen Centre as station coaches working with the children alongside the librarian.

A good example of this idea is the recent display complementing the senior students' classroom study of plants. The Allen Centre featured bottle-garden making out of recycled bottles. Bottle-garden science was presented as an interactive puzzle. The bottles were planted with Venus Fly Traps, which needed to be fed, then were linked to an insect bank created out of cardboard milk cartons and explored with a digital microscope. Also featured were two hands-on Zen Gardens with associated philosophy; an easel-full of Georgia' O'Keefe's flower paintings linked to an online gallery; two other associated workshops including a woodworking opportunity to make a Zen garden rake and a Pinning Insects Workshop where as librarian I gave input as the "expert entomologist" (the name the students gave me). The Allen Centre's Wiki space offered links to take interested students further into the study.

The Wiki, http://allencentre.wikispaces.com, our version of the library home page, has a very specific place in supporting these workshops especially because it enables students to access them on or off-site. The Wiki provides step-by-step instructions so workshop activities can be undertaken at home. It serves as an electronic resource of ideas and a visual record of engagement useful in the evidence of effectiveness required by New Zealand's Education Review Office when they visit. [*As a matter of fact, every teacher-librarian should use this resource as they advocate for their library/learning commons*—Editor's note.] The Wiki also provides an avenue for communicating with parents so they are familiar with the learning commons activities. It offers convenient access to booklists, homework sites, and interesting links—therefore defining the Allen Centre as an information hub or center "without walls" operating 24/7.

Realizing the potential of these displays is a fundamental challenge. It requires a great deal of genuine engagement through one-on-one communication between the students and "librarians," who must be receptive to questions, respond to original thinking, and be prepared to embrace unfamiliar workshop practicalities, triggered by the display. Having sparked enthusiasm, a very real part of this fireworks package is the provision of further opportunities for inquiry in different guises.

There are also opportunities for children with passions and expertise to become leaders in different areas. Recently we have had students 'run' workshops for other children and even support me running sessions for teachers/students at an ICT conference.

HOW DO WE MANAGE?

To implement our program we are staffed by a full-time librarian position, which is an unusual situation in New Zealand Primary Schools where the library is usually managed by a part-time librarian or a parent-volunteer. In The Allen Centre we have divided this full-time position into three roles: a librarian, library assistant, and an activities coordinator.

The library assistant manages the collection, monitors transactions, and trains "student-librarians." We also employ an activities coordinator who works closely with the librarian to design and implement displays and conduct workshops. In addition we have attracted a range of volunteer experts, passionate about their interests, including a geologist, a retired mathematician and musician, a creative embroiderer, a retired wood and metal-worker, and a parent-journalist with a keen interest in Lego-Logo robotics. These people are invaluable in complementing the staff's skill-set, enabling us to offer a wide range of high-quality activities.

One of the key aspects to doing our jobs is getting to know the children well—their interests, passions, reading ability, and their trails-already-followed—taking the time to discover what makes them tick and thinking of them as more than issue histories. Many of the projects we do are because we have specific students in mind and one of our strengths is that we can tailor our activities and even our book collection to what we know interest our students.

Engagement is the key. The Allen Centre staff members develop perspectives of the children that are uniquely different from the perspectives of their parents and the teaching staff. The challenge is in sharing this knowledge to benefit the children's education and wider lives.

Each year the library staff provides a comprehensive report on the topics explored and activities undertaken to the Board of Trustees and community. While we have a very high trust model for our learning commons this report also provides a summative assessment and reflection on the year's programs.

A TYPICAL DAY IN THE ALLEN CENTRE?

There is no time for cups of tea! We never factored in the time implications of a triggered child! The sessions in the learning commons are *very* active. The librarian introduces the new activity in the first 15 minutes of a session then interacts with the students by helping them find appropriate books. Book-child-matching is a huge responsibility and if you get it wrong you have lost them, but if you get it right they keep coming back! Interspersed with this is connecting with their enthusiasm; for example, responding to their enthusiasm about something they brought in for the museum as they share their discovery under the microscope. Library staff engages with them as they interact with the displays and this continues throughout the day; which this semester has included Russian at lunchtime on Fridays, robots on Thursday, music and shadows on Monday and Tuesday, sketching on Wednesdays, and Geology in between! Basically you cannot trigger a child and then pull back.

"One of the key aspects to doing our jobs is getting to know the children well - their interests, passions, reading ability, and their trails-already-followed - taking the time to discover what makes them tick and think of them as more than issue histories."

TO SEE THE TRANSFORMATION

We gave careful consideration to the internal environment of the spaces such as acoustics, daylight, artificial light, ventilation, and heating, and we are very happy with the results.

As the photographs and Wiki show, a great deal of glass has been incorporated into the buildings for natural light with floor-to-ceiling windows framing the attractive outside landscape to become part of the inside décor. A wall of bi-fold glass doors expand the inside space, allowing activities to spill over outside. Richly colored carpet and drapes provide a backdrop for high quality wooden shelving helping to make The Allen Centre a quiet and relaxing space. The workstation and furniture are mobile to allow space flexibility. Stage lighting enhances the value of the stage space for performances by students and visiting groups; and heat pumps keep the new buildings warm.

We would love to have you come visit us. We are near the south tip of New Zealand close to those fabulous mountains where *Lord of the Rings* (Jackson, 2001) was filmed. So, just hop on the plane and spend the day with us as we work with the children. However, just in case that is not possible, visit our blog at http://blog.core-ed.net/greg/2009/07/a-lunchtime-in-the-allen-centre.html.

We just think it is all about the children and what they want to learn as much as it is what we are asking them to learn.

REFERENCE

Jackson, P., Osborne, B. M., et al. (Producers), & Jackson, P. (Director). (2001). *Lord of the Rings*. [Film]. (Available from New Line Cinema/Time Warner, 888 7th Avenue, 19th Fl., New York, NY 10106.)

Greg Carroll is a principal who embraces the concept of the library/learning commons. The virtual equivalent has grown because of him. He is highly regarded in the New Zealand education community for his creative use of Internet and digital technologies as effective teaching and learning tools in the primary school environment. Carroll recently completed a Ministry-initiated (government), year-long engagement visiting South Island Primary Schools in an advisory capacity. He may be contacted at *gregc@outram.school.nz*.

Peggy Stedman is the librarian at the Outram School. She has designed, facilitated, and implemented The Allen Centre library program since its inception in 2003. She has also had considerable influence and input into the school vision and direction. She may be contacted at *peggys@outram.school.nz*.

FEATURE ARTICLE

From Book Museum to Learning Commons:
Riding the Transformation Train

"The bottom line is that transformation is possible with a strong mission, vision, and goals."

CHRISTINA A. BENTHEIM

The school library at three-year-old D.L. "Dusty" Dickens Elementary School in North Las Vegas is like Grand Central Station. It is busy. It is crowded. It is sometimes messy. It is definitely noisy. And I like it that way. When walking through the halls of the school, one can hear the library buzzing with activity from down the hall.

As teachers come in and out, they see children working at the media stations, playing board games, browsing with crumpled shelf markers in tow, sitting against a soft pillow or stuffed animal while reading independently or with a friend, writing book reviews, shelving or circulating books, practicing reader's theatre scripts, setting up displays, enjoying virtual author visits, or simply browsing the stacks. That is during circulation time. During the instructional period, students work in groups on structured media literacy activities while as teacher-librarian I am providing direct supportive reading instruction to students who are far below grade level.

As I close in on my first year as teacher-librarian, I see and hear that the library has come a long way in a very short time. This time last year the library was a sterile room where every book was perfectly placed on the shelves and there was very little reading, learning, or access for students or teachers. I learned that teachers even inquired about eliminating the librarian's position because it was so ineffective.

THE TRANSFORMATION

Newly hired three months before the 2009-2010 school year began, I focused immediately on listening, researching, and planning. The school has 797 students in a very transient (nearly 50%), high-poverty, home foreclosure-prone area. Sixty percent of the library's collection had never been checked out when I arrived (see Figure 1 for circulation numbers). I have not imposed a check-out limit on patrons. Choice continues to be a critical part of reading success and it is unethical for me to restrict book checkouts based on reading levels or quantities.

Scheduling

I have tried to incorporate as much of a flexible schedule in the library as possible even though the district only provides elementary teacher-librarians for prep relief. Students have open access to the library as early as 8'oclock in the morning to as late as 4 o'clock p.m. when the after-school tutoring program ends. Students are also free to come into the library to work, use the computers, check books in or out, get help with assignments, and more, at any time during the school day. The library is closed for only 15 minutes when I am required to perform lunch duty in the cafeteria.

The Physical Space

Fortunately, the physical space of the Dickens LMC is quite large. In addition to a spacious main area of the library, which includes a stage and viewing area, vaulted ceilings with floating kites, floor to ceiling

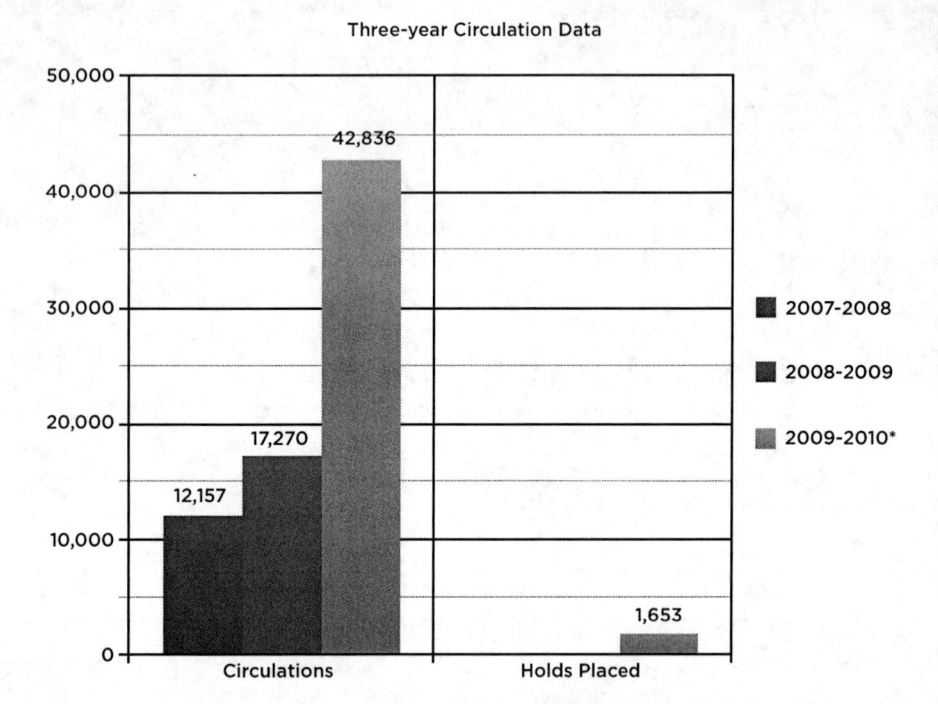

Figure 1. The LMC's circulation numbers through February 2010.

Three-year Circulation Data: Circulations — 2007-2008: 12,157; 2008-2009: 17,270; 2009-2010*: 42,836. Holds Placed: 1,653.

A first grade boy shows two other first grade students how to access a web site.

windows, and a U-shaped circulation area adjacent to the Dewey section, there are also two wings. One wing is used for the fiction and general book sections. The other wing is used for professional development and station work. In the station, tables and chairs are arranged to allow for maximum movement and learning. The media station (a bank of seven computers) is also close by, but far enough away from the other stations so that students are not distracted.

Instructional Model

The model for the instructional part of the library program was borne from recent research on school improvement, academic achievement, and reading instruction. As a former 6th grade reading teacher, I felt that a strong instructional component was needed within any program to maximize its effectiveness. With rigid reading programs fulfilling mandates, there is not a lot of flexibility for classroom teachers to teach literacy skills using authentic experiences.

Therefore at Dickens each class is divided into seven groups with approximately six to eight students per group, depending on class size, rotating at a station to focus on thematic word study, writing, media literacy, author study, reader's theatre, games, and guided reading.

In the thematic word study station, students explore the etymology of words and related concepts through games, sorting activities, web image searches, and presentations. The goal of this station is to help students learn words in contexts relevant to topics studied in their classrooms or within the library.

The writing station has a variety of authentic writing activities such as creating advertisements, responding to poems and photos, reflecting on life experiences, explaining higher-level thinking through scenarios, cultivating a writer's notebook, writing songs and plays, and more. Students engage in all stages of the writing process—including publishing—through the use of traditional and social media technologies.

At the media literacy station, students participate in webquests, research information through databases, explore Web 2.0 tools from the library web site, and write book reviews for posting within our OPAC. Assignments such as the evaluation of web sources and tools, or step-by-step research guidance using the Big 6 model, generally help students transfer their learning from the library to the traditional classroom.

The author study station provides students with text sets by a particular author, as well as biographical information so students can expand their knowledge. During an author study, students read books and compare and contrast them, write letters to the authors, create their own endings or versions of stories, and so on.

The reader's theatre station is a place where students can strengthen their reading fluency while working together as they read silly stories and poems.

In the games station, students practice their critical thinking, problem solving, and social skills through the use of games like Scrabble, Wits and Wagers, and Say

5 minutes	5 minutes	25 minutes	5 minutes	10 minutes
Check in	Lesson introduction, announcements, and book talk	Instructional literacy centers	Hands-on instruction in stacks	Check out

Figure 2. Breakdown of Instructional Period (50 minutes once per week)

A first grader cozy and comfortable while reading.

Anything.

Finally, the guided reading station is an intensive group that I lead, which helps students with reading strategies such as thinking aloud, predicting, and summarizing. We read texts and talk about them to help facilitate metacognition, which often does not happen in basal-level instruction in the traditional classroom.

While these groups are heterogeneous, I find it important to stretch the thinking skills of the gifted students so they are not always in the position of having to help less capable students. The gifted students are on a special team that works together to design the reading garden that we plan to build just outside the library windows.

Technology—The Web Site

It was during the summer that I started establishing a web presence for what is now known as the Dickens LMC (www.mrsbentheim.com). The Web was the first place where I started to create the Dickens LMC brand—one of exploration, warmth, openness, and creativity. As the brand took shape, I began emailing the school's teachers

Intermediate students developing their fluency through reader's theatre.

and administrators to share relevant links about the new LMC, seek insight via surveys, as well as educate them on my philosophy, policies, and procedures. By the first back-to-school staff meeting, teachers had a solid idea of what was to come from the Dickens LMC. In fact, they greeted me with whoops, hollers, and a partial standing ovation! I felt like a celebrity and realized this was an environment that desperately wanted the change I was setting out to provide.

WHAT THE PATRONS SAY

Teachers frequently praise the programs and services of the Dickens LMC. One teacher says, "The librarian reads to students, talks about books and technology, and lets students explore and live the library. As a teacher, I have experienced much more assistance with finding the right books and with reading programs, standards, and collaboration of lessons taught." Another teacher reports, "Students are engaged in activities, can check out books they are interested in, have learned how to use the databases, can request and reserve books, and are allowed to speak and interact." According to a staff survey I conducted in December 2009, 100% of staff respondents said they think the LMC's instructional program will greatly affect the school's test scores—and the lifelong learning of students—this year.

I also have a group of parent volunteers who help shelve, work our fundraisers, donate books and stuffed animals, and so on. The administration says the new Dickens LMC program breathed life into the school.

THE FUTURE

We have come so far, but there is still a lot of work to do as the cultivation of an effective school library program is an on-going process. In the coming years I plan to continue building the library collection with e-readers and MP3 players, design a permanent reading garden, acquire at least two Wii gaming systems, start media education programs for parents, get kindergarteners into the library to check out and read books with their parents, add a set of 40 wireless netbooks so each student has access to a com-

Fifth graders participating in a Book Pass to explore new library materials

puter in class, expand our games station to include Legos, launch a podcasting station, fully document the collaboration occurring between classrooms and the library using our Collaborative Learning Launchpad, http://dickenslmc.wikispaces.com, and inspire teachers to actively use our collaboration Ning, http://dickenslmccollaboration.ning.com, among other things. Ultimately I would like to hire a paraprofessional (we currently have a part-time assistant) that is capable of guiding students in their quest for knowledge and let the Friends of the Library Club handle most of the clerical duties such as shelving and filing.

I have been fortunate that I have administrators who are willing to let me run with my ideas, parents who are receptive, and teachers who want to be involved. However, the things going against me are plentiful too including a transient population and remaining stigma of the previous library environment.

The bottom line is that transformation is possible with a strong mission, vision, and goals. If you want change to happen badly enough, you have to be the change agent, the advocate, the publicity specialist, and the teacher-librarian, every day and to every patron.

Christina A. Bentheim is a teacher-librarian in North Las Vegas, Nevada. She received a Master's of Education in Curriculum and Instruction (School Library Science) from the ALA-accredited University of Nevada, Las Vegas. Visit the Dickens LMC at http://www.mrsbentheim.com and read the Dickens LMC blog at http://mrsbentheim.edublogs.org. She may be reached at cbentheim@interact.ccsd.net.

Part III:

Curriculum and the Learning Commons

As readers of the previous sections have already discovered, the Learning Commons concept is only as good as the program elements that happen either in the physical space or virtually. The central feature of the Learning Commons program in this section is that the entire school embraces the best practices of 21st century teaching and learning.

While there is a place for direct instruction and very sequential learning, including experiences of drill and practice, the central focus of the Learning Commons is active collaborative inquiry, problem solving, critical thinking and other strategies that promote the engagement of learners. The Learning Commons is not just a place to bring all learners up to a minimal level of performance; it is a place to push beyond minimums toward excellence.

The articles in this section draw our attention to various kinds of learners and curricular initiatives likely to succeed in a Learning Commons space. Looking at the patterns across authors and articles, we recognize a push beyond mediocrity; the creation of excitement, involvement, collaboration, and student-centered programs. There are no directives here ~~to sit down~~, nothing to read or listen to quietly, nothing to do with worksheets and then regurgitate on tests. Rather, these articles empower you to create a shared and customized vision which will address the needs of your school library, making it possible for it to become a Learning Commons.

Whether in physical or virtual space, the best educational practices thrive with teacher librarians, teacher technologists, and other adult specialists in the school as they are poised to collaborate and co-teach with the lonely and stressed-out classroom teacher. The sense is that the classroom teacher need not shoulder the entire responsibility for educating the particular students under their charge. And, reaching out for a collaborative partner is no sign of weakness or inferiority. Readers should note that the constant pattern is that two heads are better than one.

FEATURE ARTICLE

Curriculum, the Library/Learning Commons, and Teacher-librarians:
Myths and Realities in the Second Decade

In the United States, we are at the juncture between the Bush's No Child Left Behind initiative and Obama's Race to the Top with its promised billions connected to "innovation."

DAVID LOERTSCHER

As the second decade of the 21st century opens, every classroom teacher and adult learning specialist in the school (including teacher-librarians) is trying to answer the question of what needs to be taught to a generation of young people who face incredible global competition.

These same young people drop out of school at an alarming rate (Fox News, 2009), making them, along with many others, unhappy with the ways of present-day education.

In the United States, we are at the juncture between the Bush's No Child Left Behind initiative and Obama's Race to the Top with its promised billions connected to "innovation." Where are we now? What are our prospects for the future? To which wagon should teacher-librarians hitch our team? At this fork in the road, which path should we choose? What questions need exploration? What should our focus be? Perhaps a reality check linked to some possibilities might stir our thinking a bit as we face this decade.

Myth #1: What we teach in the classroom and in the library/learning commons is our prerogative.

Reality #1: Others have more and more control over what we teach, if not how we teach.

In Figure 1, try to think of all the pressures on classroom teachers and teacher-librarians that determine what we teach. Frankly, it is a conglomerate of conflicting ideas and mandates driven by assessment, competition, and expectations. During the standards era of the 1990s, many professional organizations were funded by the United States federal government to develop standards that would help teachers determine what to teach in history, language arts, mathematics, science, and other disciplines (NCTM, 2001; NCSS, 1994; AAAS, 1991; NCTE, 1996). Of course, every professional organization thought their discipline to be foundational in education so when Robert Marzano (1999) did a major review of the total pool of ideas that young people should master, the list would require 20 years to teach at a normal pace.

As a result of the major national curriculum documents, most states of the United States wrote or revised their state standards. Those documents or their revisions are in place in many schools and districts as mandates of what teachers and teacher-librarians should teach.

Then along came the elephant in the room: assessment. One would think that assessments would be built to match the mandates of the state standards documents. Not so. Across the nation, government officials sought to get a handle on "academic achievement" so they could compare across schools, districts, states, and nations. The assessments did not match the state standards neither could they be compared across the nation. There were various competing tests that cost taxpayers enormous

Figure 1. The pressures on classroom teachers and teacher-librarians that determine they teach affect curriculum in so many ways!

sums as they became very popular with the general public who wrung their hands in dismay as test score after test score did not show any improvement in what they believed it should.

In order to compare the United States to other nations whose governments instituted a single test as a yardstick, the United States federal government created the NAEP test (http://nces.ed.gov/nationsreportcard) given to a random sample of students in reading and mathematics. The results of these tests have been used in the past decade to compare student achievement internationally.

Students in the United States have not made the hoped for gains predicted that the No Child Left Behind initiative was pushing. Figure 2 illustrates this struggle between testing and the curriculum.

It was about the same time in Japan as their example of "exam hell" began to percolate into the classroom as teachers prepped students for testing rather than learning. In the United States, all eyes were focused on the classroom teacher. If only the teacher would use direct teaching techniques and cover the material to be tested, all was supposed to work. In some districts, teachers were told to be on the same page in the textbook as their colleagues across the nation so that students could theoretically get the same lesson at any school they happened to be attending that day. Thus far, the emphasis on direct teaching is not showing up on national tests. Teaching to the test does raise scores minimally in the short term but researchers have yet to demonstrate that substantial progress in either math or reading is being made. Instead, researchers report minimal to flat results. Multiple analyses of the NAEP scores exist for both math and reading. One such interpretation is by Kevin Carey (2009). What began as a good idea that no child would be left behind ended up as a program to bring every child up to a minimal level and to grade level; to fill gaps and in the end, to mediocrity.

Now, the National Governors Conference has sponsored a set of national common core standards that has been in draft form and should emerge in 2010 (http://www.corestandards.org). Reading and mathematics are the first of these standards and their effect on states and local standards are a matter of conjecture at this point.

Implication #1: Teacher-librarians in schools where textbook lecture approaches, direct teaching, and lockstep coverage are the predominant practice will find a very high wall and locked door into the kingdom of the classroom. Stimulating interest in various topics and having resources that will be about these topics and accessible to students for individual work, will not happen. This is true for any of the specialists in the school such as reading teachers, instructional supervisors, and teacher technologists. Holding on to a library program/learning commons is viewed as an expensive frill that does not and cannot raise achievement in spite of the Lance research to the contrary (2005, 2002). When budget woes hit a school, any frill program is subject to cuts immediately, particularly one as expensive as the library/learning commons.

Myth #2: Assessment is going away.

Reality #2: Assessment is here to stay for the foreseeable future.

While legislatures and governments seem unable to resist the pressures of testing, testing, and more testing in spite of its astronomical cost to the taxpayer, some modifications may be in the wings. Strong arguments have been made to test not only what a young person knows but also their 21st century learning-how-to-learn skills. While the United States currently uses the

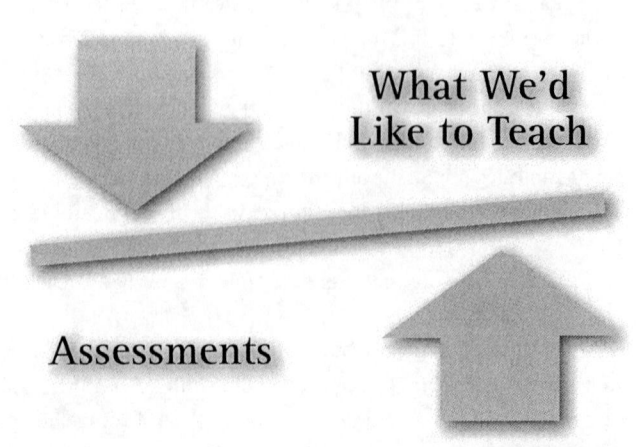

Figure 2. The current assessments model presents a weight on our best intentions.

> "What began as a good idea that no child would be left behind ended up as a program to bring every child up to a minimal level and to grade level; to fill gaps and in the end, to mediocrity."

NAEP test for international comparisons, its multiple choice mostly content-knowledge test items do not reflect higher-order thinking or other 21st century skills. One possibility is to move to the PISA international test (Programme for International Student Assessment, www.pisa.oecd.org) that requires students to read longer passages with problems that need solutions. A test like this does tap many 21st century skills so both content and process knowledge is being assessed.

Implication #2: Should the opportunity arise for multiple assessments of both content knowledge and 21st century skills, the door to teacher-librarians should be open considerably further than at any time over the last decade. Teacher-librarians will need to take advantage of such opportunities. To get prepared to do this, teacher-librarians need to assess the effect on student learning of the information and technological resources they supply to students during a learning experience in the library/learning commons. It does little good to say, "I taught them to evaluate web sites" or "They used a wiki to accomplish a collaborative task." One must be able to point to the result of better information or effect on wikified projects. What was better? For what percentage of the students? Utilize Big Think strategies (Loertscher, Koechlin, & Zwaan, 2009) to jointly assess content and process knowledge on co-taught units with teachers. If you can begin reporting outcome every time there is an intervention of the library into a learning experience, you will be ready to take advantage of any shift in assessment from content knowledge toward 21st century skills.

Myth #3: The curriculum of the classroom teacher and the learning commons are two separate spheres of work. That is, the teacher should concentrate on content knowledge and the teacher-librarian should concentrate on 21st century skills (including information literacy).

Reality #3: No one in education seems to have recognized that teacher-librarians "took over" information literacy and now 21st century skills. Both content and learning-how-to-learn has always and continues to be the role of the classroom teacher.

A popular scenario for teacher-librarians over the past two decades has been to homestead the territory of information literacy skills, create a curriculum, and teach the various skills at an appropriate grade level—all in an isolated approach. The theory has been that if one could dictate information literacy as a part of the state curriculum and test the skills, teacher-librarians would have job security.

A second scenario is that 21st century skills are means to ends and should be integrated into content learning. In this view, the teacher-librarian and the classroom teacher join forces to push both content and learning skills together. For example, during a research paper project, the real function of the research skills lesson is to teach the student how to master content knowledge. Figure 3 demonstrates this idea.

Notice that the learning skills drive the construction of deep understanding. If separated and taught in isolation, the result will be the same as pushing on the clutch of an automobile. If the clutch is disengaged, the auto goes nowhere. One can argue and find supportive research that content knowledge will never reach excellence without engaging process skills. Thus, reading skills, information skills, media literacy, critical thinking, and creative thinking are inseparably connected to content mastery and invention of new knowledge.

If a learner uses 21st century skills to master topical knowledge, we advocate that transfer and sophistication grows topic by topic across grade levels until a sense of confidence that "I can learn anything I want to learn" is part of a learner's dispositions.

Figure 3. The wheels turn together....

Teacher-librarians are sharply divided on this idea as demonstrated in two AASL publications. For example, the new national guidelines, *Empowering Learners: Guidelines for School Library Media Programs* (AASL, 2009) charges teacher-librarians to integrate information literacy into co-taught learning experiences. However, *Standards for the 21st-Century Learner in Action* (AASL, 2009) created by a different AASL committee advocates grade by grade teaching of information literacy skills as a curriculum with only some reference to integrated teaching.

Implication #3: Both sides of the argument for a 21st century curriculum taught in isolation and the integrationists need to do a major test of both systems—particularly in the age of Web 2.0 technologies to test a variety to benefits to learners using either or both ways. A major study of best practices needs to be done on this issue. In the meantime, integrationists doing Big Think metacognitive activities can document with the classroom teacher the idea that "two heads are better than one." The isolated teaching of curriculum folks should be able to document the effect of their work on the basis of learning outcomes rather than merely time and skills taught.

Myth #4: Best practices research completed in the last several decades applies to the teaching and learning of the digital generation.

Reality #4: Best practices literature often relies on a body of research literature. One of the most quoted authors in this area is Robert Marzano. If one examines the body of research on which he bases his major recommendations, we find that almost all of the research cited is before 2000 (Marzano, Pickering, & Pollock, 2001). This means the best practices are based on research with young people before the major revolution in information and technology happened. The reality for the digital generation and best practices that reach this generation of learners is still evolving (Bowman & Lackie, 2009).

Implications #4: Since many teacher-librarians have embraced Web 2.0 tools in their teaching and collaborative activities with classroom teachers, action research can help sort out what best practices of the past combined with solid strategies in the new world of information and technology actually contribute to learning and push excellence. During and at the conclusion of collaborative learning activities with classroom teachers, take the time to include both formative and summative assessments of both content knowledge and process skills. A body of local action research results will inform the adult partners about what works best with the crop of learners you have at the present time (Moss & Brookhart, 2009). Know the research and contribute to it with your own findings as well.

Myth #5: Inquiry as a teaching and learning strategy is dead in the face of state and national assessment practices.

Reality #5: Major voices across education continue to push creativity in teaching approach: inquiry, engagement, real learning, global outreach by students, project-based learning, Understanding by Design, differentiation, dispositions, and excellence. Attend any ASCD or ISTE conference. Follow professional publications of Solution Tree, Corwin Press, or Stenhouse among others to note the number of titles devoted to more constructivist strategies and ideas. There is a war of ideas and practices out there. Assessment tends to drive the behaviorist and direct teaching ideas, but teachers looking for engagement, creativity, critical thinking, and outreach are not deterred from sneaking in what they know learners want and regard more with increased effort.

Implications #5: The more teacher-librarians hitch their wagon to these counter voices, the more likely they are to lead the way toward excellence. As a profession, we have no future in direct teaching; teaching to the test; rigid objectives—lecture, guided practice, and standardized testing. The learner automatically regurgitating prescribed content is not anything that appeals to us. However, when we promote high level learning experiences in place of bird units, we must assess the difference in outcome and results, making sure we shout out the achievement of excellence. Everyone understands that when we collaborate, learners win. Those classroom teachers who ignore us do so at their learner's peril.

Myth #6: Technology is making a difference in teaching and learning.

Reality #6: Some applications do make a difference. Most do not—at least it is yet to show up in the evidence. The problem at this point seems to be that many teachers are merely transferring already poor or just plain bad assignments from one medium to another. What was poor on paper is still poor no matter the delivery system. Still others invest their time in the glamour of technology concentrating on a tech tool and its potential uses vs. starting with a learning problem and diagnosing which tech tool can help.

In the first approach, we may provide teachers with many ideas of how a Wiki can be used in their classes. In the learning to technology approach we may diagnose that a class lacks motivation to engage, so we introduce an exciting new technology coupled with a "real" learning experience. Figure 4 illustrates some of the major learning outcomes begging for tech assistance.

While we have covered many specific applications that boost specific learning challenges elsewhere (Marcoux & Loertscher, 2009), several examples here might illustrate the point of prescription to meet a learning need. For example, to increase learning efficiency in less time, we use Wikis to facilitate collaborative writing in a shorter amount of time than if products were edited serially by group members on paper. In order to build deep understanding of major concepts we might use collaborative online graphic organizers edited in real time. For inclusion of all learners, we might offer materials in text, audio, video, and assistive devices at various levels of complexity. We hold the technology accountable to produce the results we desire rather than dazzle and hope for a possible impact.

Implication #6: Teacher-librarians who ground themselves in the possible applications of technology devices, software, Web 2.0 apps, and online resources, transform that knowledge into prescriptive "cures" during collaborative lesson creation with

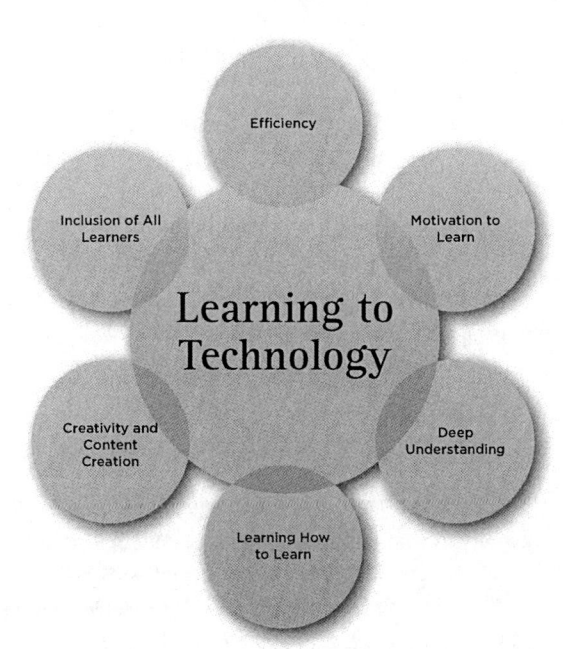

Figure 4. Technology would certainly make a difference in meeting any of these learning goals.

classroom teachers. Together the team chooses technologies to meet a learning challenge and rejoice in the results or the next time around, find a technology that will accomplish the learning goals. Classroom teachers keep coming back to the "physician" who delivers the cures.

Myth #7: School libraries make a difference in teaching and learning.

Reality #7: Many research studies done by Lance, Todd, Achterman, and others have been conducted in the following states and Canadian Provinces including Colorado, Pennsylvania, New Mexico, Alaska, Florida, Ohio, Delaware, Michigan, Minnesota, Indiana, Illinois, North Carolina, Texas, Wisconsin, California, Ontario (Canada), and Idaho. (Reports, presentations, and brochures are available at http://www.lrs.org/impact.php#on. Printed copies of studies are available at http://lmcsource.com.) Repeatedly, school libraries are linked with achievement. However, the problem lies in the type of research performed in these studies. The studies use either correlational or qualitative research, neither of which is acceptable by the United States Department of Education as gold standards research such as experimental or quasi-experimental research design.

What this means is that we cannot say certain characteristics of a school library program CAUSE higher achievement. We do know that schools that care enough to have high quality library programs also have higher achievement scores. It does not mean, if a school principal hires a teacher-librarian and implements a library program, scores will go up automatically. We might expect some influence. We might anticipate a consequence if we provide a certain set of services. Over and over and over, we do see an effect. Why? And why, if the correlation is so strong do teacher-librarians continue to lose their jobs to support personnel? In times of financial exigency, few jobs are safe, but teacher-librarians can be confident that high quality school library programs have and continue to have an effect. Experimental and quasi-experimental studies need to be done. In the same breath, both co-relational and qualitative research methods have been acceptable for a century and are likely to continue as quite acceptable research methods.

Implication #7: The various national and state studies of school library impact are interesting and informative when establishing a set of program services likely to make a difference. The problem is, however, what makes a difference in your school, with your teachers, with your set of students and community? There are a number of professional resources available that help teacher-librarians discover their own effect on teaching and learning in their particular schools (Loertscher & Todd, 2003).

Ross Todd and others have been urging and demonstrating the power of evidence-based practice on what we decide to emphasize in our programs (2004). We can be certain that support and supply services in and of themselves do not make enough difference to brag about. It is the demonstration of actual impact of technology, information, books, 21st century skill integration, and collaborative construction of learning experiences that can be probed for direct impact. Using the measures we already know day in and day out, adding up the resulting effect over time, and reporting the results to everyone who will listen is the strategy that will provide the best predictor of our indispensability in each school where we practice.

QUESTIONS FOR THE DECADE AND COROLLARIES FOR INDIVIDUAL PROGRESS

As we think of the next decade, what are the possibilities to consider? What are the questions that come to mind? Here are a few.

1. Do learners really thrive and excel in a technology-rich environment? Corollary: Do my learners actually thrive in the environment I have provided?

2. Do learners subjected to a high-quality information environment develop a higher quality of deep understanding than kids who merely surf the Net? Corollary: When I co-design learning experiences with teachers, what evidence is there that high-quality information makes a difference in their products and learning?

3. Does the move toward a client-side based learning commons place the traditional school library at the center of teaching and learning? Corollary: As my program moves toward a learning commons, what evidence do I have that it has moved closer to the center of teaching and learning?

4. Does the development of knowledge building centers that turn directive assignments into conversations among classroom

teachers, students, teacher-librarians, and other specialists have an effect on what is learned and how it is learned? Corollary: What evidence has come from knowledge building centers I have constructed that both individual and group learning was enhanced by this collaborative space?

5. Is inquiry equal to or superior to direct teaching in terms of both content understanding and 21st century learning abilities? Corollary: When the teacher and I have turned boring research into exciting inquiry, what major differences make the added effort worth the work and time invested?

6. Can teacher-librarians really develop a track record of evidence-based practice that will hold up under school, district, state, and national scrutiny? Corollary: What measures do I take on a regular basis that demonstrate results rather than the input of things, materials, teaching time, visits, and access? Who believes the data I collect and report?

Finally, looking ahead at this new decade, we can be sure of a few things:

- Technology and its potential to educate and learn will continue to grow over time.
- Alternatives to the printed book are likely to lead to more and more reading on line. It will be the kids and teens who decide what media they prefer. It will be our job to provide access to what they want and need on whatever devices they prefer and certainly wherever and whenever they care to access what we have to offer.
- The continuing global competition is not going away. High expectations for content knowledge linked with extremely strong 21st century skills is likely to predominate as we try to help our kids push beyond mediocrity toward excellence.
- Research, both formal and action research must blossom more than ever if we are to be a dynamic part of the best teaching and learning.
- High expectations for teacher-librarians in what they know and can do are rising. We will look to our best and brightest to lead this profession into the center of teaching and learning.

REFERENCES

American Association of School Librarians. (2009). *Empowering learners: Guidelines for school library media programs.* Chicago, IL: American Library Association.

American Association of School Librarians. (2009). *Standards for the 21st-century learner in action.* Chicago, IL: American Library Association.

Bowman, V. & Lackie, R., eds. (2009). *Teaching generation M: A handbook for librarians and educators.* New York: Neal-Schuman.

Carey, K. (2009). NAEP Math 2009: What it all means. Retrieved January 2, 2010 from http://www.quickanded.com/2009/10/naep-math-2009-what-it-all-means.html.

High school graduation rates plummet below 50 percent in some U.S. cities. Retrieved December 31, 2009 from http://www.foxnews.com/story/0,2933,344190,00.html.

Lance, K. C., Rodney, M. J., & Hamilton-Pennell, C. (2005). Powerful libraries make powerful learners: The Illinois study. Canton, IL: Illinois School Library Media Association. Available at www.islma.org/resources.htm.

Lance, K. C. & Loertscher, D. V. (2002). *Powering achievement—School library media programs make a difference: The evidence mounts.* 2nd ed. San Jose, CA: Hi Willow Research and Publishing.

Loertscher, D. V., Koechlin, C., & Zwaan, S. (2009). *The big think: Nine metacognitive strategies that make the end just the beginning of learning.* Salt Lake City, UT: Hi Willow Research & Publishing.

Marcoux, E., Loertscher, D. V. (2009). *Achieving teaching and learning excellence with technology. Teacher Librarian, 37*(2), 14-22.

Marzano, R., Pickering, D., & Pollock, J. (2001). *Classroom instruction that works: Research-based strategies for increasing student achievement.* Alexandria, VA: ASCD.

Marzano, R. J., Kendall, J. S., & Gaddy, B. B. (1999). Deciding on 'essential knowledge'. *Education Week*, 18(32), 68, 49.

Moss, C. M., & Brookhart, S. M. *Advancing formative assessment in every classroom.* Alexandria, VA: ASCD.

National Council for Social Studies Task Force. (1994). National Task Force for Social Studies Standards Curriculum Standards for Social Studies. Waldorf, MD: NCSS Publications.

National Council of Teachers of English and International Reading Association. (1996). *Standards for the English Language Arts.* Retrieved December 31, 2009 from http://www.ncte.org/library/NCTEFiles/Resources/Books/Sample/StandardsDoc.pdf.

National Council of Teachers of Mathematics (2001). *Principles and standards for school mathematics.* Retrieved December 31, 2009 from http://standards.nctm.org.

Rutherford, F. J., & Ahlgren, A. (1991). *Science for all Americans: American Association for the Advancement of Science project.* Cary, NC: Oxford University Press.

Todd, R. (2004). Knowing and showing how school library programs help students learn. Retrieved January 2, 2010 from http://www.accessola.com/osla/toolkit/Home/EBPLinks.html.

David V. Loertscher is coeditor of *Teacher Librarian*, author, international consultant, and professor at the School of Library and Information Science, San Jose, CA. He is also president of Hi Willow Research and Publishing and a past president of AASL.

FEATURE ARTICLE

the library is the place: knowledge and thinking, thinking and knowledge

JUDGING BY WHAT OUR SCHOOLS HOLD AS PRIORITY, WITH CONTENT KNOWLEDGE FAR SUPERSEDING THINKING SKILLS IN TERMS OF WHAT WE TEACH AND ASSESS, IT APPEARS KNOWLEDGE IS VASTLY MORE IMPORTANT THAN THINKING. BUT IS IT? CAN THEY EVEN BE SEPARATED?

If we ask which came first—knowledge or thinking—we land in the chicken-and-egg dilemma. For the two exist in an endless cycle, a feedback loop in which the *process* of thinking creates the *product* of knowledge, which informs further thinking, which creates new knowledge, ad infinitum.

Our tests, in keeping with our instruction, are almost exclusively content-based. Even when schools do try to teach thinking processes, they separate these from the content curriculum. Yet the way our brains function is to use the process in order to understand the product. We must infuse thinking skills into the content curriculum. Think of content as "what to know" and thinking as "how to know." We must approach the two together and help students understand how they work in tandem. We must teach both *what to know* and *how to know*.

LIBRARY AS LOCUS OF LEARNING

Because of this content-driven approach, classrooms can be regimented and scheduled. By contrast, the library offers students greater freedom to explore thinking through content. Teacher-librarians are uniquely placed to extend and enrich the learning process launched by teachers from their discrete classrooms.

A library is a learning commons and a place of active research. It is a locus of exploration. There, students move beyond single text provided for classroom use and into a boundless territory of many and varied resources. The library is also the place where many aspects of the students' schooling intersect: while there working on a product, they engage in and become more aware of the process aspect of their education. For this reason, the library may be the ideal location for launching a new paradigm of teaching both *what to know* and *how to know*.

We take it for granted the importance of training students in methods of research. In other words, we teach them how to go about gathering content knowledge. Librarians willingly give much time to this. What's missing is the analogous training to help them use—and hone—their thinking skills.

FIGURE 1

Knowledge

by derek cabrera and laura colosi

We do not help them think about their thinking. By and large, American students at all grade levels have no idea how to navigate cognition and learning. As teacher-librarians, as teachers, and as parents, we must be explicit with children about how the process of thinking works. Right now, we give content knowledge its due. We must do the same with thinking. Let us undertake this task in the library, where process meets product in a more pronounced way.

THE PATTERNS OF THINKING METHOD

We have been remiss in teaching thinking skills only because we have not known how. We understand knowledge because it's visible and tangible. That's why, for the most part, we're good at teaching content. But thinking is the invisible, intangible process behind that product. It's the slippery fish that we cannot grasp. How can we teach what we cannot grasp?

To understand how we think, we need only look at the tangible, visible product of thinking—knowledge—and the patterns in how it is structured. We use the visibility of knowledge to understand the invisible process of thinking. In other words, the structure of knowledge holds the patterns for how to create more of it, which is what we do in the process of thinking.

Remarkably simple and accessible, the four patterns lend themselves effortlessly to a practical, extensible method for infusing thinking skills into the curriculum—and for helping students understand their own thinking processes. The four patterns are symmetrical, each made up of two elements that complement each other:

• Distinctions: identity and other. Whenever we make a distinction, we assign ideas an identity and in so doing create an invisible other. Making a distinction about who is included in the idea of "us," for example, also creates the distinction of who is excluded and considered "them." We use distinction-making to name and define, to compare and contrast, in order to find similarities and differences, to draw and test boundaries, and to make choices.

• Systems: part and whole. Every idea is a whole system made up of parts, while also serving as a part of a larger system. We organize part-whole relationships by either splitting an idea apart into the smaller ideas that make it up or combining it with other ideas to create a new one; we often use both of these processes simultaneously. Whenever we sort ideas, nest them, or use or create categories, we're organizing systems of parts and wholes.

• Relationships: cause and effect. When two ideas relate to each other, they have a mutual effect on each other that changes them both. Recognizing relationships among ideas leads to interdisciplinarity, transfer of learning, analogical thinking, and the ability to form new ideas by combining seemingly disparate pieces of prior knowledge. Relationships are often implicit and require us to recognize them; when we do, we make associations, interactions, and connections explicit.

• Perspectives: point and view. Every idea is a perspective comprised of a point and a view. The point is the subject, or the position from which the idea is viewed; the view is the object, or what is viewed. Because the point affects the view and vice versa, we expand what we know about any idea in profound ways when we become conscious of both point and view. Making perspectives explicit increases creativity, innovation, conflict resolution skills, and prosocial behaviors such as compassion and empathy.

We could not find a single instant in the vast realm of human knowledge that did not hold the same four repeated patterns. Mathematics, physics, chemistry, biology, psychology, economics, dentistry, trail building, road paving, skateboarding, and hopscotch—nothing deviates from this structure. The four patterns of thinking are universal and ubiquitous, repeating across cultures and disciplines. All humans think this way.

THINKING ABOUT THINKING

In any teachable moment, we are often literally *one question away* from helping students gain deeper understanding of a topic. The gap between taking in a fact (which research has shown students are likely to forget) and absorbing a true understanding (which they are likely to retain)

> "Teacher-librarians are uniquely placed to extend and enrich the learning process launched by teachers from their discrete classrooms."

> "In other words, the structure of knowledge holds the patterns for how to create more of it, which is what we do in the process of thinking."

> "Finally, we can teach both what to know and how to know."

can be bridged with the right question at the right moment (Bransford and Stein, 1993; National Research Council, 2000).

Many teacher-librarians are already skilled in questioning students, especially to help them assess the research aspect of their work. They look back with them in culmination activities and consider how a project unfolded and what it led to. Such metacognitive "debriefing" events unify students as a group of learners while helping individuals know themselves as students and developing researchers.

Any number of basic questions can open the door to metacognition:

• What do we all know together about this topic?
• How did we come to know this?
• So what?
• How could we take this deeper?
• How can we improve the research process?

This simple and effective approach to teaching students to look at their research process—in other words, to understand how they gather content—can be replicated through analogous questions that enable them to track and understand how they think. The missing questions are well within our reach, falling into the four categories of the four patterns of thinking. And while they are basic and painless, they are also powerful: one simple question can bridge the gap between *what to know* and *how to know*.

The beautiful thing about the universality of the four patterns is that teacher-librarians are all already using them. The patterns are familiar. When children name and define a concept, they are making a distinction. Whenever children see cause-and-effect at play, they are recognizing a relationship. They use the four patterns daily in school and life with no idea that they are doing so. Librarians can train students in this process by getting them thinking about thinking—about how they approach any knowledge to come to understand it.

The question "How is a spider different from an insect?" trains students to consciously make distinctions. They learn to compare and contrast when they are trying to work out the boundaries of one concept in relation to similar or related concepts. Teacher-librarians can further make students aware of their thinking by being explicit about what we observe when we see them using any of the four patterns. For example, "That is an interesting differentiation you are making. Let us recognize that you are differentiating between two things, and that differentiation is a skill—a skill that falls under distinction making."

When we see students organizing part-whole systems (or working to understand how anything works), we can simply ask, "What are the parts of a fire engine?" Or of the legislative branch of the U.S. government, or of a fox den, or of absolutely any concept—all objects and ideas are made up of parts. With systems we can always help students notice that those parts have parts, and these parts have parts too. Looking the other way, we can ask, "What whole is our whole a part of?" A fire engine is part of the world of the firehouse, and that is part of our city's public safety program. Thus, students come to look for the embeddedness of part-whole structure in everything through a two-part inquiry habit: *What are the parts of X?* and *What is X a part of?*

When students need to recognize a relationship, we can ask them "How is the Gila Woodpecker related to the Saguaro Cactus?" Or, "How is the concept of *interdependence* related to the concept of *ecosystem*?" When we put two concepts together, what are we implying? When Barack Obama was a candidate for the United States presidency, an implicit relationship was made by those who published pictures of him in a turban. We help students make that relationship explicit by asking about the implications of an image that puts together Obama and a turban. "What does this relationship say about Obama?" "What does this relationship say about Islam?" This naturally leads to questions that probe content deeper: "Is this true?" "Is it valid?"

With perspective taking, we can help students understand that a perspective exists even when it is unstated. A textbook from Virginia contained the sentence, "Life among the Negroes in Virginia in slavery time was generally happy" (Simkins, Jones, & Poole, 1973). We can ask them, "Whose perspective does this idea come from?" "How would this be presented from a slave's perspective?" This example in particular brings home the fact that deeper thinking occurs for students who learn to ask themselves such questions. Another important and profound direction to go with perspective taking, especially after students have researched what others say about a topic, lies in the simple question, "What do *you* think about this topic?"

With the Patterns of Thinking method, we can stop wondering about how to teach students to think and get on with the process. For students of all ages, across all grades, through any subject matter, with any book in the library, and any enrichment activity created by teacher-librarians or generated in collaboration with classroom teachers, we can rely on the four patterns of thinking plus simple lines of inquiry to facilitate the balance between content and thinking. Finally, we can teach both *what to know* and *how to know*.

TACTILE MANIPULATIVES

Laboratory and field studies show that incorporating tactile activities into any lesson engages multiple sensory systems and areas of the brain in the learning process. Students have increased recall of information, deeper understanding of lesson content, and core thinking skills are developed (Minogue & Jones, 2006). Learners of all ages can better grasp concepts when they can literally *grasp* them with their hands.

To illustrate, consider ThinkBlocks, used by teachers, professionals, and students from pre-K to PhD to externalize and manipulate their ideas.

ThinkBlocks not only clarify the students' thinking but render the thinking process visible. ThinkBlocks support both cognition and metacognition.

Each dry-erasable ThinkBlock functions as a single idea. When we label a block *Westward Expansion*, for example, it holds that identity and becomes differentiated from every other idea. We thus make a **distinction**. Since ThinkBlocks come in three sizes, we can organize nested **systems** (Figure 2). We drop medium-sized blocks into the big block of *Westward Expansion* to represent the populations that shaped and were shaped by this historical era: *the United States government, Native Americans,*

ThinkBlocks

> While teaching a class at Cornell University to guide PhD candidates from various fields in writing their dissertations, I saw students struggle as they tried to focus, organize, and develop their thoughts. Their thinking was muddled and their ideas entangled. Knowing the importance of touch to thinking, and with the patterns of thinking in mind, I headed to my garage (where else for good solutions?). There, I sawed up some white boards and organized them into 3-dimensional blocks that the students could both handle and write on. I made more blocks in the same shape at two smaller levels of scale. Then Velcro did the job of connecting one block to another. Finally, I added a reflective surface on one side of the blocks. These were the first prototypes of ThinkBlocks. For more information, visit www.ThinkandThrive.com. —*Derek Cabrera*

"Documented research tells us that kinesthetic learners benefit highly from the gesture and movement involved in handling blocks and ddropping them into one another."

immigrants, and *the Lewis and Clark expedition.* We can dig deeper by exploring the characteristics and beliefs held by each of these populations. We do this by dropping small blocks into the medium blocks and naming those traits—one trait per block.

FIGURE 2

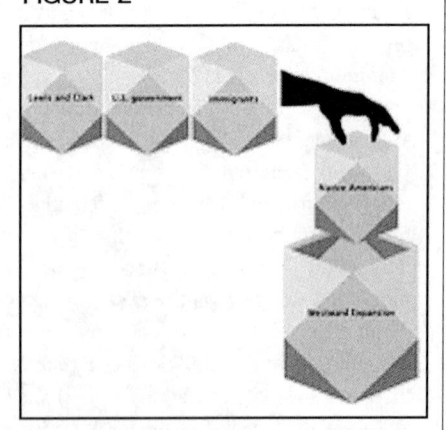

Westward Expansion

The key is in handling the blocks and dropping them into other blocks as we name the part of a larger whole that each one represents. Documented research tells us that kinesthetic learners benefit highly from the gesture and movement involved in handling blocks and dropping them into one another (Pfeifer et al., 2006; Striano et al., 2003).

Because ThinkBlocks are magnetic, we can construct the **relationship** between the land and the four populations we identified as characters of the Westward Expansion. We do this by connecting a ThinkBlock labeled *Native American* to a ThinkBlock labeled *Land* (Figure 3). If we insert a smaller block between the two, that block represents the **relationship**.

Students can "blow up" and zero in on the **relationship** by using a large block and then building it as a **system**. We drop in medium blocks as we name the different

FIGURE 3

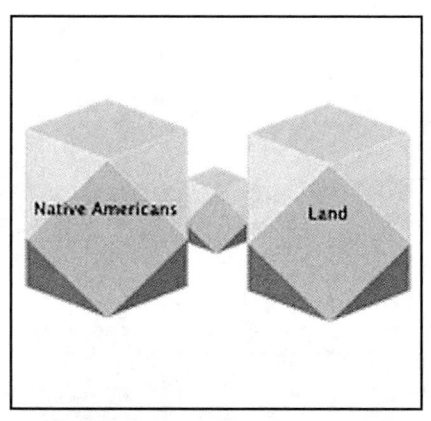

Native Americans

aspects of the **relationship**: they lived close to the land; they lived on land without owning it; they had already been there for many generations. It is important not only to identify that a **relationship** between two ideas exists but, more important, to explicitly examine the **relationship** as a separate idea in and of itself that can be further examined as a distinct **system**.

Turning the reflective surface of one ThinkBlock toward another demonstrates **perspective-taking**. In Figure 4, we take the *Native American* block and point its reflective surface toward the *Westward Expansion* block and ask, "How did the Westward Expansion look to the Native Americans? What did it mean to them? How did they respond?" Students see that each population involved has a different **perspective** on the event.

FIGURE 4

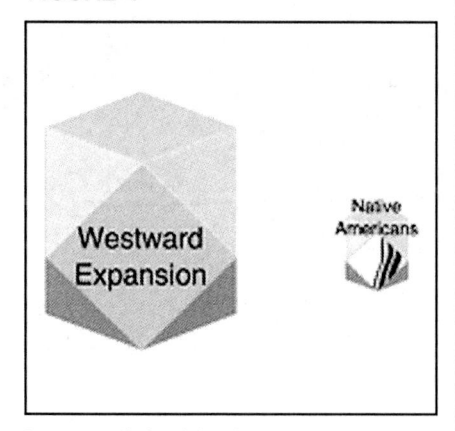

Image of the block

The students' thinking skills are strengthened as they learn to consider any event from more than one point of view.

As they do this repeatedly throughout their education, they learn that any reporting of an event is shaped by a certain perspective—and that we must recognize the influence of perspective on the potential accuracy of any account.

ThinkBlocks are not necessary for a deep and thorough use of the four patterns of thinking. They are useful because they nest and connect and have surfaces you can write on plus one reflective surface. But it is possible to work physically with the patterns of thinking by using bowls and buttons or pieces of chicken and toothpicks. You just cannot write on pieces of chicken.

INTEGRATING THE FOUR PATTERNS

However you work with them, the four patterns of thinking work together to form an integral whole. None exists singly in isolation; all four are at play in all that we think about. We cannot make a single distinction without eliciting part-whole thinking, relational thinking, and perspective taking.

As children get older, they can integrate and become aware of integrating the four patterns. They can go beyond formulating a distinction to reformulating the distinction by taking a different perspective. Teacher-librarians can build off the thinking skills children acquire in the lower grades by taking their questioning a step further into integration. For example, once we recognize a relationship, what are the parts of that relationship? How do the parts of a system change if we look at it from another perspective?

This gets increasingly sophisticated in the upper grades. If we have a system whose parts we have defined, such as biology, and we relate it to another system, such as chemistry, what are the results? Biochemistry is not only the relationship between biology and chemistry. It becomes a whole system unto itself with its own name (representing a new distinction made) and its own unique parts—books, web sites, conferences, experts, teachers—all particular to biochemistry.

Throughout high school, as students prepare for college, we can train them to apply the patterns of thinking to any specific content knowledge and any area of research. This will enable them to achieve their full academic potential in any subject area.

CREATING 21ST CENTURY THINKERS

Students schooled with the four patterns of thinking infused into their curriculum will be able to make distinctions, organize part-whole systems, recognize relationships, and take multiple perspectives. They will also have a number of crucial skills in place that integrate the four patterns of thinking. All of this amounts to the building of robust thinking skills. Students will be able to do the following:

- Identify and distinguish the parts of a relationship.
- Sort, group, nest, or categorize ideas from many different perspectives.
- Recognize the part-whole structure of Distinctions, Systems, Relationships, and Perspectives
- Reformulate a distinction by taking the perspective of the "other".
- Recognize and note the invisible "other" whenever new parts, perspectives, or relationships are formed.
- Demonstrate relational thinking by taking second and nth order perspectives.

The mastery of those skills should occur progressively throughout their education. Then, what we will see as they graduate and move out into the world is that they have become 21st Century thinkers with advanced thinking skills and the ability to solve complex problems and meet new knowledge without fear. They will be able to:

- Differentiate between the content and the structural patterns of ideas.
- Recognize like structural patterns in unlike content.
- Recognize similar structural patterns across subject areas to facilitate interdisciplinary transfer.
- Build analogies, metaphors, and similes and demonstrate how small changes transform meaning.
- Innovate new solutions to complex problems by seeing alternative constructions and avoiding lock-in.
- Demonstrate flexible thinking skills to adapt to changing needs or variables.

> As teacher-librarians continue to use their unique opportunity to be flexible with our learners, they can emphasize the beauty and richness of the content in every book and activity even as they consistently teach students to think. And just as they teach them to look at their research process along the way, they can equally teach them to think about their thinking process as they move through the process of acquiring a broad base of knowledge.

Reflection Question: Why four patterns of thinking?

• Construct simple, elegant, universal, fractal, symmetrical models.

As teacher-librarians continue to use their unique opportunity to be flexible with our learners, they can emphasize the beauty and richness of the content in every book and activity even as they consistently teach students to think. And just as they teach them to look at their research process along the way, they can equally teach them to think about their thinking process as they move through the process of acquiring a broad base of knowledge.

REFERENCES

Bransford, J. D. &. Stein, B. S. (1993). *The Ideal Problem Solver: A Guide to Improving Thinking, Learning, and Creativity*, 2nd ed. New York: Freeman.

Minogue, J. & Jones, M. G. (2006). "Haptics in Education: Exploring an Untapped Sensory Modality." *Review of Educational Research* 76(3): 317-48.

National Research Council. (2000). *How People Learn: Brain, Mind, Experience, and School*. Washington, D.C.: National Academy Press.

Pfeifer, R., Bongard, J., Grand, S., & Brooks, R. (2006). *How the Body Shapes the Way We Think: A New View of Intelligence*. Cambridge, Massachusetts: MIT Press.

Simkins, F. B., Jones, S. H., & Poole, S. P. (1957). *Virginia: History, Government, Geography*. New York: Scribner.

Striano, T., Rochat, P., & Legerstee, M. (2003). "The Role of Modeling and Request Type on Symbolic Comprehension of Objects and Gestures in Young Children." *Journal of Child Language* 30(1): 27-45.

Derek Cabrera, who holds a PhD from Cornell University, is the founder and president of ThinkWorks. The company's vision is Thinking at Every Desk. He is a senior faculty member of the Research Institute for Thinking in Education and a research fellow at the Santa Fe Institute for the Study of Complex Systems. He can be reached at *dac66@cornell.edu*.

Laura Colosi, PhD, holds a senior faculty appointment at Cornell University's Family Life Development Center. She has more than fifteen years of research and teaching experience at Cornell University in the area of parenting as it relates to developing children who are better thinkers and learners. Laura is the vice president of ThinkWorks. She can be reached at *ergo@ThinkandThrive.com*.

FEATURE ARTICLE

The Role of a School Library in a School's Reading Program

"...the connection between the reading program of the school library and reading in a K-12 school is most often non-existent."

ELIZABETH "BETTY" MARCOUX AND DAVID V. LOERTSCHER

As a foundational element of schooling, learning to read and reading to learn is every bit as important in the 21st century as it has ever been. Whether interacting online, doing assignments, taking tests, or social networking, the ability to read and read well affects every part of our existence and often predicts success throughout life.

WHAT IS OUR CURRENT ROLE?

As editors of *Teacher Librarian*, we have seen the development of major national initiatives dealing with reading both in the United States and in Canada over the past several decades. We have been concerned as we read professional literature and research about reading that the connection between the reading program of the school library and reading in a K-12 school is most often non-existent.

What program connections can the professional teacher-librarian make that contribute to the success of the reading program in the school? In other words, besides being a warehouse of reading materials upon which everyone draws, what initiatives spring from collaboration between the classroom teacher, the reading specialist, and the teacher-librarian in a school that pushes foundational literacy?

We began with a variety of foundational documents that formulate a role checklist for libraries at the heart of a reading program and then asked the *Teacher Librarian* advisory board and other professionals to respond. Today, as you build a learning commons model for your school library, here is our collaboratively built checklist for consideration. Which of the following program elements exist in your school to increase reading? Which of these are ones you can take on as a programmatic goal this year in your school library/learning commons? Are there missing elements that are contributing to reading in your school?

ACCESS TO PROFESSIONALS AND SUPPORT PERSONNEL

• All Pre-K–12 classroom teachers are knowledgeable in building reading skills.
• Full time reading specialists serve all students in building reading skills and fluency.
• Full time credentialed teacher-librarians are knowledgeable in the teaching of reading and the development of life-long readers.
• Administrators have background education in reading across the school and prioritize reading for all students.
• Full time support personnel handle the warehousing of physical reading materials, access, circulation, and repair, making reading accessible to all while supporting the professionals' reading habits and behaviors.
• All faculty and staff model life-long reading importance to all students.

> "Whether interacting online, doing assignments, taking tests, or social networking, the ability to read and read well affects every part of our existence and often predicts success through life."

> "The teacher-librarian serves as the building's reading advisor on materials for the reading community."

• Physical and virtual collections of professional resources are built to support reading.

• Face to face and virtual discussion forums are created to build capacity collectively.

ACCESS TO BOOKS AND OTHER READING MATERIALS

• There is access to a plethora of reading materials below, at, and above one's reading level with unlimited circulation to classrooms and homes.

• Circulation policies allow and encourage children to check out a variety of self-selected books to read regardless of "level."

• Learners are provided with their choice of many genres and formats both fiction and informational.

• Access to materials in print, digital, audio, and in any combination includes access to the technologies necessary to use these materials.

• Every student and teacher has a preferred device(s) on which to enjoy reading in any format.

• Every student and teacher receives information about new and innovative reading opportunities from the teacher-librarian.

• Regardless of language, there is appropriate access to a wide variety of multicultural materials for the reading community.

• Access to fictional resources crosses a wide range of interest levels and genres.

• Access to informational resources crosses a wide range of disciplines and personal interests.

• Continuous selection and weeding of materials and technologies provide current resources for the reading community.

• There is financial and professional compensation from the school for the lack of reading materials in a student's home.

• Rotating classroom collections from the library are readily available, frequent, and in large quantities. Every classroom is a print-rich environment.

• The entire school participates in the building of, access to, funding of, and maintenance of a bountiful collection of reading materials.

• The teacher-librarian serves as the building's reading advisor on materials for the reading community.

• Students and teachers participate in resource suggestions and thus feel ownership of library collections.

A WHOLE SCHOOL READING COMMUNITY

• A reading leadership team at every school includes the reading specialist, the teacher-librarian, administrators, and classroom teachers.

• A whole-school reading thread runs through every school improvement plan. Articulated instruction is reading focused.

• Reading is one of the major issues of the professional learning community or other school improvement initiative/structure and is openly discussed by all concerned parties.

• Reading initiatives draw in parents and the community, especially the public library community. Being a school-wide initiative, reading motivates wide participation and access—all building toward the habit of life-long reading.

• Everyone in the school is aware of the school-wide reading initiative and participates actively in it.

• Everyone in the school is using the power of technology to discuss, share, recommend, reflect on, critique, and encourage reading in the same ways they are doing in face-to-face groups. These discussions can extend beyond school boundaries and involve not only local schools, but also those from national and global areas.

• Data are collected to mark the progress of the reading program of the entire school. Data from the library's reading program folds into this assessment.

• The library media center serves as the hub of the entire reading program of the school, with a clear understanding of how to best facilitate all reading in the school.

• The value placed on reading is evident throughout the school.

INTEGRATED READING INSTRUCTION ACROSS THE GRADE LEVELS

- Reading skill, while introduced and cultivated in the early grades, remains integrated throughout all grades as needed by individual learners. Access to reading materials and cultivating reading interest are even more important than many prescriptive and repetitious reading skills.
- Sophistication of reading skills goes beyond decoding, fluency, and vocabulary instruction into understanding and critical response to difficult texts, points of view, reasoning, interpretation, analysis, and synthesis.
- If any member of the teaching or administrative staff lacks preparation to participate effectively in the whole-school reading plan, professional development is available and encouraged. This is particularly true at the middle and high school grade levels.
- If a strategy is not working with a particular student or group of students, alternative strategies are tried until something succeeds.
- If textbooks are too difficult or too easy for individual readers, the library is the source for appropriate textual materials on all levels and in a variety of formats. Differentiated instruction involves textbook understanding as well as project comprehension.
- The school's collaborative reading program extends into the development of other foundational literacies involving writing, media literacy, visual literacies, and efforts to build 21st century learning skills.
- Advice about successful reading programs is regularly sought from local, state, and national reading documents, reading research, standards, professional literature, and reading organizations.
- Action research is used to confirm what works in order to boost the entire reading program of the school, whether looking at individuals, groups, or the entire school.
- Reading is valued as a critical strategy to build vocabulary and background knowledge.
- Knowledge of the needs of different kinds of reading behaviors is gathered and applied to reading instruction and special initiatives. Example, boys and reading, enrichment, and other special needs, etc.

TIME TO READ

- Time to read is a daily part of the school day and beyond. Reading is done before and after school, but more important, reading is scheduled during school hours.
- Time to read is adjustable for individuals as well as classes and large groups, but all students, teachers, and staff read daily.
- Time to read includes both required and recreational reading. The types of reading materials available are plentiful and tailored to the needs, abilities, and interests of the individual.
- Reading aloud daily to students of all ages is part of the time-to-read strategy.

INTERACTION WITH COMMUNITY RESOURCES

- The reading leadership team promotes collaboration with the public library on its programs and reading opportunities. Particular attention is paid to public library reading initiatives during school vacations.
- The reading leadership team seeks out community organizations and volunteers that can assist and support school-wide reading initiatives.
- District, state, regional, national, and international reading organizations available most often through the Internet become part of the school-wide reading program.
- Professional development of all staff includes participation in activities that enhance school reading initiatives.
- Reading contests and extrinsic motivation are replaced with collective goals and emphasis on the idea that "reading is its own reward." This idea is modeled often in the school.
- The reading leadership team communicates on a regular basis with the community about reading initiatives and programs.

"Knowledge of the needs of different kinds of reading behaviors is gathered and applied to reading instruction and special initiatives."

"The biggest question is what can we do collaboratively that we cannot do separately?"

FAMILY SUPPORT

• Parents and siblings are involved in building a reading community at home, and become partners in school reading initiatives.

• Efforts to include the home are particularly important if poverty or cultural values make reading less central to the home environment.

• Efforts are made to provide whatever means, resources, or training the family needs to elevate reading in the home.

• Parents are encouraged to borrow materials from school and public libraries.

• The reading leadership team helps parents to understand that many things count as reading, for example magazines, graphic novels, blogs, web sites, etc.

COLLABORATIVE READING ACTIVITIES AND CELEBRATIONS

• School and public libraries collaborate to promote a plethora of reading activities.

• Activities that produce direct results replace those that take a great deal of time, efforts, and means but produce little in terms of results.

• Build a mix of strategies from a variety of reading activities (such as reading buddies, DEAR, digital book clubs, library card campaigns) that work in your educational community.

• Design interactive learning experiences with authors, illustrators, and poets.

• Develop authentic opportunities for readers to creatively express their connections to reading.

• Utilize technologies to encourage readers to discuss, write about, and communicate their understanding and build understanding with others.

CONCLUDING IDEAS

As we look across the entire educational literature, various professional groups seem to carve out their own isolated role in stimulating reading competency. Often, there is little to no communication among the various groups about what constitutes a vision of what works. This document assumes that each school has a literacy leadership team that brings them together. It represents a united effort in a school rather than fractured or competing ideas in the pursuit of reading excellence for a school's students. The biggest question is what can we do collaboratively that we cannot do separately?

Each school faces a unique set of strengths and challenges given their students, who have unique characteristics as the digital generation, a diverse cultural and language generation not found in previous generations.

The home environments from which students come also factor into their learning situations. With an attitude of continuous improvement or perpetual beta (as it is called in technology), sometimes fine tuning is in order. Research also finds that major overhauls are needed for every adult in the school regarding technology, collaboration, and student achievement. With so many student achievement expectations being set by states, provinces, or national governments, it is the school library media program that can bind these together and make things happen in all schools.

FOUNDATIONAL DOCUMENTS AND ORGANIZATIONS

AASL documents and resources

• AASL Reading4Life @your library®: Position paper on the library media specialist's role in reading. Approved January 2009.

• AASL Values: http://www.ala.org/ala/mgrps/divs/aasl/aboutaasl/aaslvalues/aaslvalues.cfm

• AASL Strategic Plan: http://www.ala.org/ala/mgrps/divs/aasl/aboutaasl/aaslgovernance/aaslofdocuments/aasl_strategic_plan.pdf

• Standards for the 21st Century Learner: http://www.ala.org/ala/mgrps/divs/aasl/guidelinesandstandards/learningstandards/standards.cfm

• Best Sites for Teaching & Learning: http://www.ala.org/ala/mgrps/divs/aasl/guidelinesandstandards/bestlist/bestwebsitestop25.cfm (includes ReadWriteThink and Thinkfinity—both associated with the International Reading Association)

• National Board for Professional Teaching Standards: http://www.nbpts.org/

• Essential Links: Resources for school library media program development: http://aasl.ala.org/essentiallinks/index.php?title=Table_of_Contents

• How school library media specialists can assist you. Reading with your children: http://www.ala.org/ala/mgrps/divs/aasl/aboutaasl/aaslcommunity/quicklinks/el/elread.cfm

• Position Statement—Resource based instruction: Role of the school library media specialist in reading development: http://www.ala.org/ala/mgrps/divs/aasl/aaslproftools/positionstatements/aaslpositionstatementresource.cfm

• Position Statement—The value of independent reading in the school library media program: http://www.ala.org/ala/mgrps/divs/aasl/aaslproftools/positionstatements/aaslpositionstatementvalueindependent.cfm

IRA Documents

• Standards for Reading Professionals (2003): http://www.reading.org/downloads/resources/545standards2003/index.html

• Category Descriptions of Reading Professionals (2007): http://www.reading.org/downloads/standards/definitions.pdf

• Adolescent Literacy: http://www.reading.org/Libraries/Position_Statements_and_Resolutions/ps1036_adolescent.sflb.ashx

• Young Adults Choices for 2008: http://www.reading.org/Publish.aspx?page=JAAL-52-3-YAChoices.html&mode=retrieve&D=10.1598/JAAL.52.3.6&F=JAAL-52-3-YAChoices.html&key=F54E706B-8EFF-4B38-84BD-2A24CA6C8A3B

• Phonics Through Shared Reading (podcast): http://www.reading.org/downloads/podcasts/CA-Gill.mp3

• Children's Choices for 2008: http://www.reading.org/Publish.aspx?page=RT-62-2-CChoices.

html&mode=retrieve&D=10.1598/
RT.62.2.8&F=RT-62-2-CChoices.
html&key=FD021889-A670-4C5E-
BFDB-40E6F2B752C7
- Teaching Reading Well: http://www.reading.org/Libraries/Reports_and_Standards/teaching_reading_well.sflb.ashx
- Prepared to Make a Difference: http://www.reading.org/Libraries/Reports_and_Standards/1061teacher_ed_com_features.sflb.ashx
- Lectura y Vida home page: http://www.lecturayvida.org.ar/
- Accreditation Redesign: http://www.ncate.org/public/062309_TeacherEdRequirements.asp

NCTE Documents

- What do we know?—summaries of current educational research: http://www.ncte.org/policy-research/wwk (just being designed)

Reading Research Reports

- IRA Supports Key Findings in Early Literacy Report: http://www.reading.org/Libraries/Press/pr_NELP_report.sflb.ashx
- Rethinking Reading Comprehension Instruction: http://www.reading.org/Publish.aspx?page=/publications/journals/rrq/current/index.html&mode=redirect
- 11 Factors That Help Schools Achieve: http://www.whatworksinschools.org/factors.cfm
- NEA to Read or Not to Read: A Question of National Consequence: http://www.nea.gov/research/ToRead.PDF

Universal Standards Information

- Report on First Draft of Common Core State Standards Initiative (CCSSI) http://www.eschoolnews.com/news/top-news/?i=59934
- Program Assessment Tool for Race to the Top: http://www.ed.gov/about/reports/annual/expectmore/index.html
- The Web-Standards Project: http://www.webstandards.org
- ISTE/NETS Standards: http://www.iste.org/AM/Template.cfm?Section=NETS
- Partnership for 21st Century Standards Skills: http://www.21stcenturyskills.org/index.php?option=com_content&task=view&id=254&Itemid=120
- 21st Century Standards Based Curriculum: http://wvde.state.wv.us/teach21/thebigpicture/21stCBigPicturePg2.htm

Canadian Connections

- OSLA Ontario School Library Association—Reading Literacy: http://www.accessola.com/osla/bins/content_page.asp?cid=750-753
- Research on School Libraries and Literacy Achievement: http://www.accessola.com/osla/bins/content_page.asp?cid=630-639
- Ontario Ministry of Education – Boys' Literacy: http://www.edu.gov.on.ca/eng/curriculum/boysliteracy.html
- Ontario Ministry of Education –Think Literacy Project: http://www.edu.gov.on.ca/eng/studentsuccess/thinkliteracy/library.html
- School Library Information Portal (SLIP)–Reading: http://www.clatoolbox.ca/slip/english/School_Library_Programs/Reading/
- Teacher-Librarians Supporting Student Learning: http://www.saskschools.ca/curr_content/teachlib/read_lit/readmain.htm
- Booth, David. *It's Critical.* Toronto: Pembroke, 2009.
- Booth, David. *Reading Doesn't Matter Any More.* Toronto: Pembroke, 2006.
- Koechlin, Carol & Zwaan, Sandi. *Building Info Smarts: How to Work with all Kinds of Information and Make it Your Own.* Toronto: Pembroke, 2008.
- Koechlin, Carol & Zwaan, Sandi. *Q Tasks: How to Teach Students to Ask Questions and Care about their Answers.* Toronto: Pembroke, 2006.

Others

- *Becoming a Community of Middle Grade Readers: A Blueprint for Indiana.* Evansville, IN: Middle Grades Reading Network, April 2009.
- Krashen, Stephen. *The Power of Reading: Insights from the Research.* 2nd Ed. Libraries Unlimited, 2004.

- Loertscher, David V. and Achterman, Douglas. *Increasing Academic Achievement through the Library Media Center: A Guide for Teachers*. Hi Willow Research & Publishing, 2003.
- Bush, Gail. *Every Student Reads: Collaboration and Reading to Learn*. AASL, 2005.
- Champlin, Connie, Loertscher, David V., and Miller, Nancy A. S. *Raise a Reader at Any Age: A Librarian's and Teacher's Toolkit for Working with Parents*. Hi Willow Research & Publishing, 2005.
- American Recovery & Reinvestment Act: http://www.recovery.gov
- Reading Rockets–Teaching Kids to Read and Helping Those Who Struggle: http://www.readingrockets.org/audience/professionals/librarians
- All About Adolescent Literacy: http://www.adlit.org
- Online School Implements Gaming: http://www.eschoolnews.com/news/top-news/index.cfm?i=59030
- Not going to college: How about a "career diploma" from high school: http://www.csmonitor.com/2009/0630/p02s18-usgn.html
- ASCD Smartbrief articles: http://www.smartbrief.com/servlet/encodeServlet?issueid=286F80BF-9764-46AE-A069-D21E44E25526&tsid=be9ebfec-1c25-4f50-829b-cd24663daf03
- Focus on data bolsters case for SIF: http://www.eschoolnews.com/news/top-news/index.cfm?i=60169
- RealNetworks loses critical ruling in RealDVD case: http://news.cnet.com/8301-1023_3-10307921-93.html
- California lists state approved digital textbooks: http://www.eschoolnews.com/news/top-news/index.cfm?i=60153
- A high school's leap from so-so to special: http://www.lasvegassun.com/news/2009/jul/27/high-schools-leap-so-so-special
- Race to the Top Guidelines announced: http://ascd.typepad.com/blog/2009/07/race-to-the-top-begins.html
- ED rules on ed-tech stimulus funds: http://www.eschoolnews.com/news/top-news/index.cfm?i=59920
- Literacy Connections: http://www.literacyconnections.com/ReadingAloud.php
- Reading Is Fundamental: www.rif.org/parents/tips
- Johnson, Doug. "Demonstrating Our Impact: Putting Numbers in Context Part 1." *Leading and Learning*, 2006-07, #2.
- Johnson, Doug. "Demonstrating Our Impact: Putting Numbers in Context Part 2." *Leading and Learning*, 2006-07, #3.
- Johnson, Doug. "Linking Libraries and Literacy: A Review of the Power of Reading." *KQWeb*. Mar/Apr 2005.
- Moreillon, Judi. *Collaborative Strategies for Teaching Reading Comprehension: Maximizing Your Impact*. ALA Editions, 2006.
- Literacy Matters: http://www.literacymatters.org/parents/ideas.htm

APPRECIATION

Our appreciation goes out to the following individuals who contributed to this document: Jack Humphrey, director of the Indiana Middle Grades Reading Project funded by the Lilly Endowment, provided the initial document which stimulated the idea; Susan Ballard, Doug Johnson, Gail Bush, Jo Ellen Misakian, Michele Farquharson, Connie Champlin, Jean Donham, Carol Koechlin, and Betty Morris contributed their ideas to this collaborative endeavor.

Elizabeth "Betty" Marcoux, a part-time faculty member of the Information School, University of Washington, is a co-editor of *Teacher Librarian*. She may be reached at *b.marcoux@verizon.net*.

David V. Loertscher is coeditor of *Teacher Librarian*, author, international consultant, and professor at the School of Library and Information Science, San Jose, CA. He is also president of Hi Willow Research and Publishing and a past president of the American Association of School Librarians. He can be reached *davidlibrarian@gmail.com*.

FEATURE ARTICLE

Influencing Positive Change:
The Vital Behaviors to Turn Schools Toward Success

"... I define a teacherpreneur as 'someone who organizes a classroom venture for learning and assumes the risk for it.'"

VICKI DAVIS

Editor's Note: For this article, we have asked Vicki Davis to envision and share her views on the "state of technology in education and the challenges to which we need to rise."

Euripides said, "Nothing has more strength than dire necessity." Clearly, this is where we are in education.

With dropout rates soaring, standardized test scores stagnant, budgets being cut, and businesses arguing that educators are not providing the skill set students need to help them be successful, we are at a turning point.

Many educators feel like the proverbial "bad child" who is always in trouble and told all the things he cannot do (don't make Johnny hate reading, don't let Suzy quit school, and stop letting them fight in the lunchroom and stream it to YouTube.) A cacophony of voices arising from such books as *Disrupting Class* (2008), *Grown Up Digital* (2008), and *The World is Flat* (2007), declares the shortcomings of education and dire consequences if we do not change.

W. Edwards Deming, father of modern industrial engineering, says "It is not enough to do your best; you must know what to do, and THEN do your best." In his book, *In Influencer: The Power to Change Anything*, change researcher Ken Patterson (2008) states:

"The breakthrough discovery of most influence geniuses is that enormous influence comes from focusing on just a few *vital behaviors*. Even the most pervasive problems will often yield to changes in a handful of high-leverage behaviors" (p. 23).

What are these high leverage behaviors? From watching the transformation of my own school and reading current research, I hypothesize that there are six vital behaviors that hold the key to the positive transformation of schools.

ENCOURAGING TEACHERPRENEURSHIP AND ACCOUNTABILITY

Research cited by the *Wall Street Journal* in a February 2008 article, *What Makes Finnish Kids So Smart*, states that Finland was named the best education system in the world. The article says:

"'Finnish teachers pick books and customize lessons as they shape students to national standards...' In Finland, the teachers are entrepreneurs".

An entrepreneur is "someone who organizes a business venture and assumes the risk for it." So, I define a teacherpreneur as "someone who organizes a classroom venture for learning and assumes the risk for it." This term can broadly be applied to anyone who works with students or organizes student learning: teachers, teacher-librarians, IT integrators, and curriculum directors. Teacherpreneurship is truly an attitude that permeates a school at every level.

Susan Israel's book *Breakthroughs in Literacy* (2009), analyzes case studies in K-8 classrooms where teachers had breakthroughs in student reading. Israel concludes her analysis with this powerful statement:

"What we learn... is that teaching is more than giving students a choice... or linking instruction with students' learning styles. It is about personalizing teaching for specific students, lessons, or skills" (p. 194).

In fact, the success of many charter schools according to the authors of *Disrupting Class* is that they give educators "the freedom to step outside the department structures...with the flexibility to create new architectures for learning" (Clayton, M. C., p. 209). As shown in these three examples, clearly teachers must customize the classroom.

When looking at organizations that successfully change, Herold and Fedor (2008) point out,

"There is no such thing as 'organizational change...'. When we say an organization has made the transition from 'point A' to 'point B,' we really mean many individuals within the organization have changed their behavior, so that collectively the organization now reflects these changes" (Herold, D. M., & Fedor, D. B., p. 70).

So, how do we empower the customization of the classroom? First, we must realize how many restrictions educators have. Educators often have no choice in their own classrooms. Their lesson plans are written for them, or even worse... scripted (Bernard, S., 2007). In many cases, they are not even allowed to arrange the classrooms in the ways they want because they have to share space with other teachers (Armstrong, C., 2009). Change leader Don Berwick, who was head of the Institute for Health Care Improvement's 10,000 Lives Campaign, said "The biggest motivators of excellence are intrinsic. They have to do with people's accountability to themselves" (Patterson, K., p. 109). The 10,000 Lives Campaign saved 10,000 lives by helping healthcare professionals make better decisions by appealing to their intrinsic motivation to do no harm. In the same way, I believe successful change will appeal to the motivation of teachers to help their students learn.

So, how do we unleash the intrinsic desire of teachers to help students learn and help teachers make the sacrifices it will take to get there? Figure 1 shows it well that "The difference between sacrifice and punishment is not the amount of pain but the amount of choice" (Patterson, K., et al., p. 106).

So, it is time we give educators choices. Hold them accountable to the standards, but let them choose the tools, web apps, web sites, resources, software, technology, and perhaps textbooks that will best help their particular class learn based upon the learning styles and unique interests of the students. Let them create spaces for learning that may not be in traditional neat rows or involve school desks at all!

Teacher engagement precedes student engagement and to engage our teachers to make the sacrifices necessary to promote excellence, we must empower teachers. Teacher-librarians, tech directors, and other specialists in the school are essential partners for teachers who are willing to change. Administrators should encourage this AND hold teachers and those who support them accountable, or teacherpreneurship will just be another failed initiative. Teachers cannot shoulder this alone.

BUILDING THE BRICKS AND CLICKS: ASSEMBLING THE TOOLS THAT FLATTEN CLASSROOMS AND EXPAND MINDS

Don Tapscott in his book *Grown up Digital* (2009) analyzes today's generation of students and recommends:

"Instead of delivering a one-size-fits-all form of education, schools should customize the education to fit each child's individual way of learning. Instead of isolating students, the schools should encourage them to collaborate" (p. 122).

This individualization and collaboration happens in two places: face-to-face and online. Both should be customized for and by the students and educators inhabiting these spaces. Perhaps no theory better embodies this thought than the Learning Commons:

"We posit that both adults and young people need to learn to build their information spaces and to learn to be responsible for their actions in those spaces. Since our clients are under our influence only part of the day, we need to help them learn and create rules of behavior in both the real world and in the digital world" (Loertscher, D., Koechlin, C., & Zwann, S., 2008, p.3).

The Learning Commons is the perfect companion to the teacherpreneur and is a space that houses librarians, media resources, technology resources, and IT integrators in a common place that is designed to be both functional and comfortable with some spaces even resembling the comfortable seating areas found in the local coffee shop. "The Learning Commons as the center of school improvement, offers a lifeline from the frustration often expressed in the teacher's lounge" (Loertscher, D., Koechlin, C., & Zwaan, S., 2008, p.65).

But learning extends beyond the school

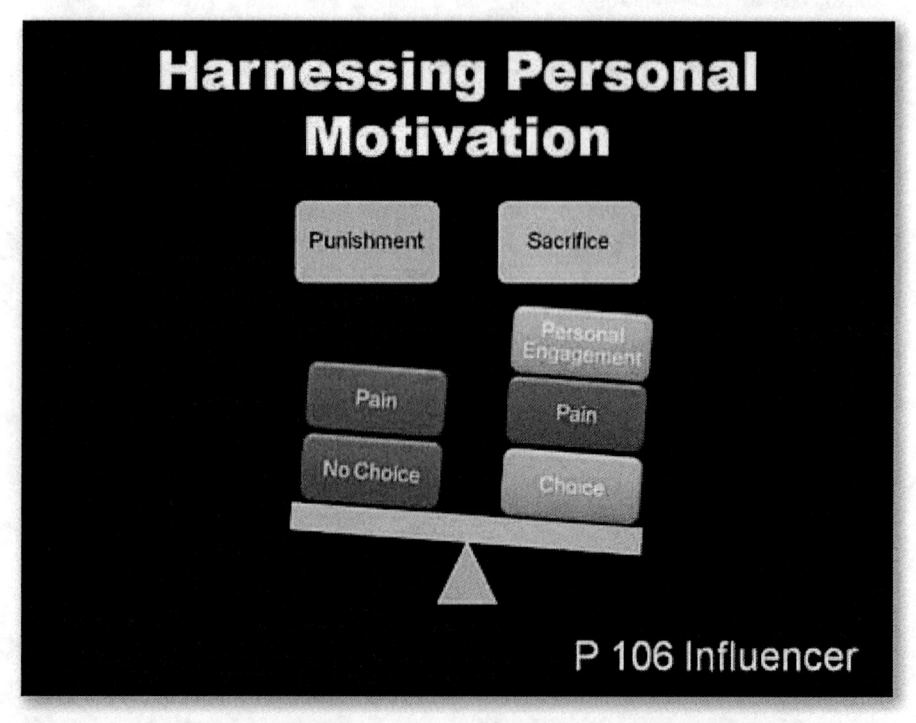

Figure 1. According to Ken Patterson, the difference between sacrifice and punishment is the amount of choice.

yard onto the Internet. As perhaps the only positive side effect of the threat of H1N1, many schools are rushing to find places online for their students to collaborate and for the first time looking at moving outside their walls. The sites a school will access are places the teacher-librarians, tech directors, administrators, and classroom teachers should discuss frequently as they explore what other schools in their area and around the nation are using. Although fear often holds many schools back, there are amazing, beneficial learning experiences using just about any Internet tool.

In addition to providing a wide variety of tools, schools like mine are looking for others around the world to "partner with a purpose" as we create common curricular projects that teach subject matter, 21st Century tools, digital citizenship, collaboration, and culture. To learn more about these global collaborative projects, visit http://www.flatclassroomproject.org. World class curriculum directors have world maps with push pins marking the global experiences of their students. World class IT directors will continually seek out new, safe tools to hone the technological prowess of their students.

KAIZEN: EMPOWERING PERSONAL LEARNING NETWORKS

"Everyone is going to need to make an audacious commitment to learning to survive" (Porter, B., p. 47).

The third vital behavior is to empower all employees to develop a personal learning network so that they may continually research new topics and refine their practice. The Japanese call this method of improvement Kaizen which means "continuous improvement" (Ten 3, 2009).

In research on the best teachers at the college level, Ken Bain (2004) found that:

"Great teachers are not simply great speakers or discussion leaders; they are more fundamentally, special kinds of scholars and thinkers, leading intellectual lives that focus on learning, both theirs and their students'" (p.134).

Currently, our model of professional development in schools is a "binge"-approach where we have educators sit in 10 to 20 hours of class over several days. This rarely creates systemic change. However, embedded professional development programs such as "23 Things" by Helene Blowers (23 Learning 2.0 Things: http://plcmcl2-things.blogspot.com/), are showing amazing, transformational change in their participants. In my own career, it was four years ago when I committed to take fifteen minutes three times a week for my own personal research and development and that practice has improved my classroom the most.

Personalized learning must begin with the adult educators in the school. We should develop these personal initiatives ourselves or perhaps alongside our students. This has happened with the Tech Angel program in New Zealand (Tapscott, 2009).

LOOKING AT PERFORMANCE AS PART OF THE PROCESS

"Assessment and individualized assistance can be interactively and interdependently woven into the content-delivery stage, rather than tacked on a test at the end of the process" (Clayton, M. C., 2008, p. 111).

I find that as I am teaching students to construct movies on a topic such as Digital Citizenship or the trends in information technology that the richest learning experiences and assessments occur at that moment. We must differentiate instruction based upon the learning styles in our classroom allowing students to record, act, reflect, blog, video, program, and engineer products that represent their learning on a topic. We must also evolve in how we assess a student's progress through a body of knowledge and mastery.

Many schools are so eager to master the test that the test has become school and that makes Jack a dull, frustrated boy uninterested in coming to school just so he can take another test. We forget that except for professional exams like the MCAT, life does not have written tests. As we move toward improving our education systems, we must also evolve and improve our assessment methodologies.

When teacher-librarians, tech directors, and other specialists in the school collaborate and co-teach, if they all adopt an assessment attitude throughout a learning experience reflecting with the students about what they know and are able to do, and then reflect together as adults, the likelihood of excellence in teaching and learning is exponential rather than incremental.

21st Century Tools

HOW MANY OF THESE TOOLS CAN YOUR TEACHERS AND LIBRARIANS ACCESS?

- Cloud Computing Tools: Google Docs, Drop.io, Zohowriter, Diigo, Delicious
- Collaborative Tools: Wikis, Google Docs
- Video Resources: Discovery Streaming, Blip.tv, Ustream, Youtube
- Educational Networking Tools: Ning, Wikis, Learn Central, TappedIn, Twitter, Edmodo, Google Apps Education Edition
- Graphic Organizing Tools: Classtools.net, Gliffy, Mindmeister
- Educational Gaming: Classtools.net, Flashcard Sites, Funbrain, Playnormous
- Mashup Tools: Firefox Add-ons, Yahoo pipes, Diigo
- Virtual Worlds: Second Life, ReactionGrid, Opensim
- Handheld Devices: iPhone/itouch Apps, Google SMS searching from cell phones

If they cannot access these tools, is there a method for teachers, librarians, and curriculum directors to request access for tools to be unblocked? If not, you are definitely NOT empowering teacherpreneurship and are stifling innovation and growth.

SELECTION OF THE RIGHT MESSENGERS

"The message is no more important than the messenger," says Donald Hopkins of the *Carter Center*, which was responsible for amazing results in eradicating the painful guinea work in 23 of 30 targeted African countries. Hopkins found that when outsiders moved into communities they were met with polite nods and very little action. It was when they began working with the chiefs and medicine men of the village that they saw improvement (Patterson et al., 2008).

Our schools have become too dependent on outside consultants and presenters while allowing change leaders and knowledgeable experts to languish in their cubicles, unnoticed and un-empowered to help things change.

Research shows that effective promoters of change spend a disproportionate amount of time with two types of leaders: formal leaders and opinion leaders (Patterson et al., 2008). Formal leaders are the administrators and those who have staff reporting to them. Opinion leaders are the people who are knowledgeable, generous with their time, and trustworthy and they often are the vital link between an entire school system and positive change.

In order to help teachers incorporate methodologies to improve student learning we must honestly look at the messengers of such change. Defining the messenger is not to be relegated to a marginal afterthought but as a paramount decision that will determine whether your initiative is adopted or becomes just another byword. Additionally, it should never be the whim of just one person on your school staff to sift through the wide variety of messages in education today but instead, teacher-librarians, tech directors, teachers, curriculum directors, and other specialists should be included in the planning school practices.

REEVALUATE DATA STREAMS

To change a person's focus, one must take a look at the data that the person focuses upon. "The fact that different groups of employees are exposed to wildly different data streams helps explain why people often have such different priorities and passions," says Patterson (p. 234). With the wide use of student information systems and the explosion of data mining, we must be careful that we are showing the proper data stream.

Patterson and others emphasizes in his book that to change behavior we must change the data stream. The thing that concerns me about the data streams in most schools is that they only consist of one thing: standardized tests. Two years ago, I heard researcher Dr. Robert McLaughlin speak on this very topic; he said,

"the assessment industry owns conversations that educators started–like math standards... We should be having educators talking with educators about what excellence looks like and how it needs to be fostered. We need to be cataloging best practices in learning technology. Our terms as professional educators should be to catalog our content. It is not hard, it just isn't happening" (Davis, 2007).

But why are we only looking at standardized testing when other research suggests that "a seldom-examined factor, student aspirations, plays an integral role in students' educational accomplishments"? (Plucker, J. A. & Russell J. Q., 1998, p. 252-257). By looking at student aspirations, student environments for learning (including incidents of violence), and other research-proven factors, we can improve the process of learning and thus improve the outcomes of learning. By the time the low test score comes back it is too late. We must refocus on the data that helps us focus on the process. But as we harness our data streams, we must be careful not to swing toward too much data as the research also shows that leaders "often undermine the influence of the data they so carefully gather by overdoing it" (Patterson, p 235).

"Teacher-librarians, tech directors, and other specialists in the school might have data streams connected with their own specific tasks such as network speed, circulation of materials, and number of lessons delivered about information literacy or networks. But, these data streams, standing alone, do not measure our effect on teaching and learning. It would be much better to select a few data streams of our own that

> "By looking at student aspirations, student environments for learning (including incidents of violence), and other research-proven factors, we can improve the process of learning and thus improve the outcomes of learning."

demonstrate our effect on teaching and learning. These are the measures that put us at the center of school improvement" (Loertscher, D., personal communication, October, 18, 2009).

IN CONCLUSION

Technology is intertwined throughout these six key vital behaviors we should encourage among educators and in schools to help facilitate change. However, it is never about the technology but about how the technology is USED to improve learning. Doing our best and trying hard is not enough if we are doing the wrong things.

Right now, the only certainty ahead of us is that we must sacrifice our time, energy, and creativity if we are to turn the course of education. And yet, the few years it takes to turn this most important institution of society are but a glimpse in the long span of education, which began when Socrates sat on a rock instructing his students orally. "We are all in this together" and this is indeed perhaps the noblest battle—the battle for success—being fought in our society today. For this is the battle for the very future of our planet and one, my fellow educators, which we cannot afford to lose.

REFERENCES

Armstrong, C. (2009). Twitter-Sig225. Retrieved October 20, 2009 from http://twitter.com/sig225/statuses/5034517577.

Bain, K. (2004). *What the best college teachers do.* United States of America: the President and Fellows of Harvard College.

Bernard, S. (2007). Edutopia poll: Do the benefits of scripted curricula outweigh the drawbacks? Retrieved October 5, 2009 from http://www.edutopia.org/node/3408/results.

Blowers, H. (2009). 23 Things. Retrieved October 20, 2009 from http://plcmcl2-things.blogspot.com/.

Clayton, M. C. (2008). *Disrupting class: How disruptive innovation will change the way the world learns.* New York: McGraw-Hill.

Davis, V. (2007, June). "Live blogging first International Leadership Summit: Dr. Robert McLaughlin's keynote." Available from http://coolcatteacher.blogspot.com/2007/06/live-blogging-first-international.html.

Gamerman, E. (2008). "What Makes Finnish Kids So Smart." *The Wall Street Journal.* March 28, 2008 http://online.wsj.com/public/article/SB120425355065601997.html.

Herold, D. M. & Fedor, D. B. (2008). *Change the way you lead change: Leadership strategies.* Stanford: Stanford University Press.

Israel, S. E. (2009). *Breakthroughs in Literacy.* San Francisco: John Wiley.

Loertscher, D. V., Koechlin, C., & Zwann, S. (2008). *The new learning commons: Where learners win!* Salt Lake City: Hi Willow Research and Publishing.

Patterson, K., et al. (2008). *Influencer: The power to change anything.* New York: McGraw-Hill.

Plucker, J. A. & Russell, J. Q. (1998). "The student aspirations survey: Assessing student effort and goals." *Educational and Psychological Measurement* (58)2, 252-257.

Porter, B. (2004). *DigiTales: The art of telling digital stories.* Sedalia: Bernajean Porter.

Tapscott, D. (2009). *Grown up digital.* New York: McGraw-Hill.

Ten3. Glossary. Retrieved October 20, 2009 from http://www.icsti.ru/rus_ten3/1000ventures_e/business_guide/glossary_lean_kaizen.html.

Vicki Davis is a teacher and IT director at Westwood Schools, Camilla, GA, author of the Cool Cat Teacher Blog, http://coolcatteacher.blogspot.com, and co-founder of the internationally recognized Flat Classroom™ projects, which have connected more than 4,000 students around the world. She may be reached at *coolcatteacher@gmail.com*.

> "Right now, the only certainty ahead of us is that we must sacrifice our time, energy, and creativity if we are to turn the course of education."

FEATURE ARTICLE

everyone wins: differentiation in the school library

DURING THE COURSE OF AN AVERAGE DAY, TEACHER-LIBRARIANS TOUCH THE LIVES OF HUNDREDS OF PEOPLE IN THEIR SCHOOL COMMUNITIES: STUDENTS, TEACHERS, ADMINISTRATORS, AND PARENTS. YET, THEY RARELY STOP TO THINK ABOUT THIS OR TAKE THE TIME TO ADD UP THE EFFECT OF THEIR WORK BECAUSE THIS IS WHAT LIBRARIANS DO: THEY ASSIST OTHERS ON THE ROAD TO LEARNING.

Furthermore, the school library functions as a learning center common to all students and staff, and its physical walls are no longer barriers to learning as school librarians and districts develop extensive support through virtual libraries that are open to patrons 24/7. Among the physical and virtual resources, technologies, and support materials, all students and teachers can find what they need when they need it.

The teacher-librarian, an über responsive educator, is the magician who makes all this happen with seeming ease. Teacher-librarians build inclusive collections, design physical and virtual work spaces for different kinds of tasks, support all students in research and independent reading quests, learn about and acquire the best technology tools, and all the while dance to the tune of individual teacher, school, district, and provincial needs. It is time to stop, reflect on, and celebrate all the ways TLs currently provide for differentiation, effective teaching, and successful learning.

The high praise of Allison Zmuda (2008) is a needed tonic for the school library world (Text Box 1). Zmuda understands that inside the school library is a curriculum specialist who can provide a distinct array of teaching strategies, a wealth of resources, technologies, strategies, and facilities that can be applied to the benefit of all teachers and learners. Classroom teachers and administrators who are struggling to address the needs of their diverse learners often overlook this obvious invaluable resource available to the whole school.

Connecting kids and content in meaningful ways is the work of all educators, and helping every child achieve is our mutual goal. However, addressing the special learning styles, intelligences, interests, needs, and abilities of every learner is the major challenge. Teacher-librarians who are so busy 'doing it' need to be reminded they are big players in ensuring that all learners have the opportunity to excel, and that as teacher-librarians they contribute to learning through the expertise they bring to the school.

Carol Ann Tomlinson (2006), a leader in the field of differentiated learning, tells us that differentiation is acknowledging that kids learn in different ways, and responding by doing something about that through curriculum and instruction. She explains that differentiating instruction is not an instructional strategy nor is it a teaching model. It is in fact a way of thinking, an approach to teaching and

TEXT BOX 1

The library media center has long been a beloved and specialized learning environment for students, a place rich with opportunities to pursue specialized inquiries, interests, and ideas. It is the most natural venue in schools for differentiation, integration of technology, and collaboration. In recent years, state and national standards for information literacy and technology have delineated a framework for what students are expected to know and be able to do as a result of their work in the library media center. Noted education researchers, system leaders, and authors as well as foundations have further bolstered the importance of the library media center as an integral part of 21st century learning so that students are prepared for the demands of the workplace. There has never been a more exciting or potentially powerful time to be a library media specialist.

Allison Zmuda. (2006) *Where Does Your Authority Come From? Empowering the Library Media Specialist as a True Partner in Student Achievement*

by carol koechlin and sandi zwaan

> *Education is about learning. Learning happens within students not to them. Learning is a process of making meaning that happens one student at a time.*
> Carol Ann Tomlinson and Jay McTighe (2006). *Integrating Differentiated Instruction and Understanding by Design.*

learning that advocates beginning where students are and designing experiences that will better help them to achieve.

In their book *Integrating Differentiated Instruction and Understanding by Design,* Tomlinson and McTighe (2006) suggest teachers first need to establish standards for student achievement and then need to design many paths of instruction to enable all learners to be successful. To reach desired learning standards, Tomlinson and McTighe encourage teachers to differentiate for students through the following design elements:

- content (what students learn and the representative materials)
- process (activities through which students make sense of key ideas using the essential skills)
- product (how students demonstrate and extend what they understand and can do as a result of a span of learning)
- learning environment (the classroom conditions that set the tone and expectations of learning).

In school libraries daily, teacher-librarians are already addressing each of these elements of differentiation. This rich potential could be spread more equitably throughout the entire school community if teacher-librarians were provided with more knowledge about differentiated instruction and if their role in contributing to student achievement was honored. It is our hope that this article will provide a starting point for teacher-librarians and initiate discussions and actions that involve the school library in a formal, more conscious role in the process of differentiating instruction.

HOW DOES THE TEACHER-LIBRARIAN CONTRIBUTE TO DIFFERENTIATED INSTRUCTION FOR ALL STUDENTS?

The school library is a learning lab, a literacy classroom, a common space for students and teachers to study, research, read, think, question, argue, discover, connect to the world, or just curl up and relax with a good book. The school library is an information and technology rich learning environment that all students should have ready access to whenever they need it. The rich diverse resources, flexible spaces, technology tools, and the expertise of the teacher-librarian provide a goldmine of assistive choices for teachers and students. It is just not possible to offer all these options to learners in an isolated classroom.

If we look at the elements of differentiation as identified by Tomlinson and examine the potential impact of the school library program we can build a pile of evidence that demonstrates the positive influence of teacher-librarians on learning for all. Everyone wins at the library. We have brainstormed a few ways that teacher-librarians differentiate in the library in order to support learning for everyone in the classroom. We hope this will be a beginning for school librarians to build on.

DIFFERENTIATION THROUGH CONTENT

The classroom teacher identifies the content to be taught and often has some classroom resources to support content learning. The teacher-librarian can enrich classroom resources and help teachers with the diverse materials and learning supports needed to ensure differentiation by:

- developing topic explorations in the library to ensure working knowledge of background information and to build related vocabulary;
- assembling resources and sources that challenge and excite but are accessible and meaningful to a range of abilities;
- ensuring that a variety of resource types (at varying levels of difficulty and sophistication), are available to support content in all subject areas, e.g. fiction, nonfiction, magazines, newspapers, databases, reference

"Teacher-librarians who are so busy 'doing it' need to be reminded they are big players in ensuring that all learners have the opportunity to excel, and that as teacher-librarians they contribute to learning through the expertise they bring to the school."

Reflection Question: What does differentiated learning look like in a Learning Commons?

"When classroom teachers partner with the teacher-librarian to design, facilitate, and evaluate learning experiences, the opportunities for differentiation are doubled."

Reflection Question: How does the school library contribute to more opportunity for differentiated instruction?

materials, websites, multimedia, and contact with experts;
- building diverse collections that reflect school demographics;
- introducing students to a wide range of reading genres to expand their reading horizons;
- connecting students to experts on topics and enhancing their ability to network with the world community;
- developing reading lists and pathfinders to support specific lessons and units;
- teaching students to examine all relevant perspectives when exploring issues.
- providing students with techniques for evaluating sources for relevance and reliability;
- helping students to be responsible users of information and ideas.

Classroom teachers can work to the benefit of many more students by implementing patterns of instruction likely to serve multiple needs.

Carol Ann Tomlinson and Jay McTighe, (2006) Integrating Differentiated Instruction and Understanding by Design.

DIFFERENTIATION THROUGH PROCESS

When classroom teachers partner with the teacher-librarian to design, facilitate, and evaluate learning experiences, the opportunities for differentiation are doubled. Together they can design and provide:
- "fail-proof" scaffolding such as topic-specific graphic organizers, question or thought prompts, student contracts and checklists, and assessment tools;
- resource-based challenges that build critical and creative thinking processes (research, problem solving, decision making, invention etc.);
- high-think learning models that align with/match the learning need (compare/contrast, jigsaw, timeline as found in *Beyond Bird Units*, Loertscher, Koechlin, and Zwaan, 2007);
- information literacy instruction infused as needed;
- a range of both student and teacher directed learning opportunities;
- instructional strategies for interdisciplinary learning;
- supportive groupings for collaborative learning experiences: discussion groups, book clubs (both face-to-face and virtual), reading/learning buddies, blogs, wikis, e-projects, and so on;
- assessment tools and strategies that help students grow.

DIFFERENTIATION THROUGH PRODUCT

The school library is arguably the most real world learning center in any school. Consequently, within the library there are many resources that will help students learn about preparing effective products and presentations. The teacher-librarian can provide opportunities for differentiating the product by:
- sharing many authentic product exemplars with students such as: posters, pamphlets, articles, video, Power-point, archived speeches, and presentations;
- introducing students to a range of effective presentation types such as press releases, photo essays, skits, interactive games, etc;
- teaching students to plan, and providing spaces for rehearsal of sharing.
- assisting students to learn about their best presentation style;
- teaching students how to use and integrate multimedia effectively in their product/presentations;
- providing authentic venues for building and sharing expertise;
- helping students extend sharing or take action beyond the school walls (e.g. letters, proposals, surveys, blogs).

DIFFERENTIATION THROUGH LEARNING ENVIRONMENT

Educators strive to ensure that every classroom provides a safe, supportive, nurturing learning environment in which students of all abilities and learning styles thrive. In all learning spaces we expect and encourage the highest possible levels of achievement

for all learners. The school library is one of those classrooms that meets these standards but also, by its very nature, contributes to more opportunity for differentiated instruction by:
- offering a real world setting for relaxing and working;
- providing multiple spaces for individual small group and whole class learning;
- matching resources to students whatever their skill level;
- creating flexible open spaces for drama, circle activities, and presentations;
- maintaining computer pods/labs with Internet access and productivity capabilities;
- acquiring simple to complex software tools;
- budgeting for multimedia tools (cameras, printers, scanners, smart boards, etc.);
- arranging quiet areas for study and relaxation (soft seating, study carrels, seminar rooms);
- designing virtual library spaces (available 24/7) for study, support, and relaxation;
- supporting students with homework help from the school library webpage;
- helping students manage their information resources and work spaces, both physical and virtual;
- providing students with self-evaluation tools and helping them set goals for improvement.

HOW DOES THE TEACHER-LIBRARIAN CONTRIBUTE TO DIFFERENTIATED INSTRUCTION FOR SPECIAL STUDENTS?

Differentiation accommodates all student needs including students with low skill levels, second language deficits, and limited background knowledge as well as gifted learners. The chart on the next two pages from *Beyond Bird Units: Thinking and Understanding in Information-Rich and Technology-Rich Environments* (2007) provides some specific examples of how teacher-librarians can differentiate for specialized student needs.

CONCLUSION

In summary we have a few suggestions for next steps. We encourage teacher-librarians to:
- **celebrate** what you already do to address the learning needs of your school community.
- **analyze** the demographics of your school population and student achievement levels so you can build even stronger, more supportive, library collections.
- **learn** all you can about differentiated instruction and related topics such as learning styles, multiple intelligences, multiple literacies, and brain-based learning.
- **build** up your toolbox of instructional strategies so you have a rich range of interventions to offer students and teachers.
- **use** the elements of differentiation as a framework for developing long range plans and budget proposals.
- **gather** the evidence of your successes as you support learning for all through your library's facilities and programs. Share your story with your administration and staff.
- Keep everyone winning at the library!

REFERENCES

Loertscher, D., Koechlin, C., & Zwaan, S. (2007). *Beyond bird units: Thinking and understanding in information-rich and technology-rich environments.* Salt Lake City, UT: Hi Willow Research and Publishing.

Tomlinson, C. A. (1999). *The differentiated classroom: Responding to the needs of all learners.* Alexandria, VA: Association for Curriculum Supervision and Development.

Tomlinson, C. A. and McTighe, J. (2006). *Integrating differentiated instruction + understanding by design.* Alexandria, VA: Association for Curriculum Supervision and Development.

Zmuda, A. (2006). Where does your authority come from? Empowering the library media specialist as a true partner in student achievement. *School Library Media Activities Monthly*, 23(1), 19-22.

Carol Koechlin and Sandi Zwaan have worked as classroom teachers, teacher-librarians, educational consultants, staff development leaders, and as instructors for Educational Librarianship courses for York University and University of Toronto. In their quest to provide teachers with strategies to make learning opportunities more meaningful, more reflective, and more successful they have led staff development sessions for teachers in both Canada and the United States.

They continue to contribute to the field of information literacy and school librarianship by coauthoring a number of books and articles for professional journals. Their work has been recognized both nationally and internationally and translated into French, German, Italian, and Chinese. They can be contacted at *koechlin@sympatico.ca* and *sandi.zwaan@sympatico.ca*.

QUICK CHECK FOR INFORMATION TASKS DIFFERENTIATION STRATEGIES

Differentiation, in the view of Tomlinson and McTighe (*Integrating Differentiation and Understanding by Design.* ASCD, 2006), requires us as teachers to set the bar of achievement for all learners but provide many paths to meet those objectives. As learning activities are created, consider adjusting them to meet the various groups listed in the table below.

	Low Skill Level	**English Language Learners**	**Limited Background Experiences**	**Gifted Above Level Skills**
Scheduling	• design flexible time with support, teacher, teacher-librarian, learning buddy • adjust task to student strengths and needs	• schedule adequate time with support, teacher, teacher-librarian, learning buddy • provide a checklist of tasks to be accomplished	• provide time and opportunity for exploring/experiencing the topic • chunk the process • provide a contract	• encourage "transfer" activities through other disciplines • help students target strengths and interests • develop an individual learning plan with negotiated timelines
Building Background	• provide visual resources and a variety of media texts at appropriate levels of difficulty • create a word wall • design activities to help connect old and new learning	• provide visual resources and a variety of media texts at appropriate levels of difficulty • plan vicarious experiences • create a vocabulary list • design activities to help connect old and new learning	• provide visual resources and a variety of media texts at appropriate levels of difficulty • provide speakers, excursions, vicarious experiences e.g. video • create topic webs • design activities to help connect old and new learning	• provide access to more complex resources • interview, survey, poll • speakers and excursions • design activities to help connect old and new learning • build a collaborative web space for sharing findings
Questioning	• encourage and model questioning • provide topic specific question starters, and focus words • use questioning aids and organizers • provide exemplars	• encourage and model questioning • experiment with question building, use manipulative cards, question starters, and focus words • provide exemplars	• encourage and model questioning • provide time to explore the topic and develop topic related concepts • experiment with question building, use manipulative cards, question starters, and focus words • provide exemplars	• encourage and model questioning • teach the Question Matrix, Bloom, and de Bono • experiment with question building manipulative cards, question starters and focus words • students collect exemplars
Collecting, Notetaking, Organizing Data	• model each strategy • provide graphic organizers, templates • use visualization techniques • use sticky notes • use technology assists	• model each strategy • provide photocopies for highlighting • provide graphic organizers, template • use visualization techniques • use sticky notes	• model each strategy • share a variety of types of templates and explain their benefits • use visualization techniques • use sticky notes	• model a variety of strategies • use software to design and create organizers and templates • use visualization techniques • use sticky notes
Visualizing Information	• share a collection of visual samples in a variety of forms • provide sample templates specific to need • work in a team	• share a collection of visual samples in a variety of forms • provide sample templates specific to need • work in a team	• provide time to explore the topic and develop topic related vocabulary • provide a collection of visual samples in a variety of forms • provide sample templates specific to need • work in a team	• utilize software for creating and presenting visualizations • work as a team or lead a team

Reproduced with permission from Loertscher, Koechlin, and Zwaan. *Beyond Bird Units: Thinking and Understanding in Information-Rich and Technology-Rich Environments.* Hi Willow Research and Publishing, 2007.

CHART CONTINUED

	Low Skill Level	English Language Learners	Limited Background Experiences	Gifted Above Level Skills
Working with Information	• model the strategy and work with students • pair with a more skilled learning buddy • create mixed skill groups with attention to skills and role assignments • physically manipulate data (stickies, index cards) • use graphic organizers to sort data	• model the strategy, work through several examples • partner with a peer with stronger language skills • allow time up front to explore the topic and develop subject related vocabulary • physically manipulate data (stickies, index cards) • use graphic organizers to sort data • utilize technologies to store, manipulate, and present data	• model the strategy and work through several examples • allow time at the beginning to explore the topic and develop subject related vocabulary • physically manipulate data (stickies, index cards) • use graphic organizers to sort data • provide thinking cues • collaborate to test ideas • dramatize information	• review possible approaches • group with similarly skilled students • act as learning facilitators or coaches • work independently and collaborate with others to test ideas • utilize technologies to store, manipulate, and present data • create graphic organizers to sort and analyze data
Establishing Criteria for comparisons or decision making	• model with concrete objects • establish criteria for sorting with the class • guided practice	• model with concrete objects • establish criteria for sorting with the class • explain criteria using visuals if possible • guided practice	• examine jobs and real life problems requiring this skill • use illustrations and video to provide background for criteria selection • work with a small group to narrow down criteria	• identify jobs and real life problems requiring this skill • establish and justify criteria independently • work with a small group to target criteria • design a matrix
Summarizing	• provide photocopies for students to highlight main ideas and supporting details in different colors • utilize video to provide background building information • summarize orally to a peer • model with a think aloud	• work with a skilled reader to identify main ideas and supporting details • create a summary with help from a learning buddy • utilize video to provide vocabulary and background building information • summarize orally to a peer • practice paraphrasing to peers	• utilize video to provide vocabulary and background building information • provide photocopies for students to highlight main ideas and supporting details in different colors • summarize orally in small groups • practice paraphrasing peer summary	• utilize technologies to store and manipulate summaries • compare summaries in small groups, look for patterns and examine discrepancies • work with other students to help create summaries of more complicated material (ESL or lower skill level students)
Thinking: So What? What Next?	• provide question prompts • scaffold discussions • design achievable thinking activities rather than large projects • share findings, build charts, webs, diagrams, maps... to reinforce	• provide question prompts • scaffold discussions • utilize word wall and vocabulary list to enhance and facilitate discussion and reflection • share findings, build charts, webs, diagrams, maps... to reinforce	• pair or group with enthusiastic students who attempt next steps • provide examples • create visuals to demonstrate learning	• make links to online quests, • discover others who have taken amazing next steps, see **www.Knowville.org** • take action • extend learning • share with the wider community
Reflection & Goal setting	• conference with each student throughout the task • peer conferencing one on one • learning log	• conference with each student throughout the task • peer conferencing one on one • learning log • utilize communication technologies to talk about learning	• conference with each student throughout the task • peer conferencing one on one or in small groups • learning log • utilize communication technologies to talk about learning	• conference with each student throughout the task • peer conferencing one on one or in small groups • learning log or reflective journal • utilize communication technologies to talk about learning

65

Cultivating Curious Minds:
Teaching for Innovation through Open-Inquiry Learning

JEAN SAUSELE KNODT

"Children feel connected to a time and place that affirms their choices and builds on the curiosity they have to offer."

> *Creativity now is as important in education as literacy, and we should treat it with the same status.* —Robinson, 2006

Today, perhaps more than ever, the world is looking for people with innovative spirits, who seek and take on new challenges, develop rich thinking processes, and visualize unique possibilities. There is a need for individuals with flexible critical and creative thinking throughout all vocations and professions. There is a need for people who are geared toward collaboration, team risk taking, and finding ways to offer unique contributions. The question to ask of our schools and parents is: *How are you preparing children to meet their futures—our futures—with innovative minds?*

Children's innovative thinking sets sail when the natural inquisitiveness they bring to the learning table is inspired, affirmed, and cultivated. When given the opportunity to openly ask and explore, children learn and thrive.

What better place than in libraries to set the stage and build industrious thinking communities? With its many resources and abilities to interface and circulate ideas, the library is an ideal place to bring folks together, establish a centralized *open*-inquiry lab for a school, and to cultivate curious, innovative, and contributing minds.

This article describes the action and possibilities of a centralized open-inquiry learning lab focused on teaching thinking to all students (K-6) as part of their regular school schedule. All quotes from lab students are presented as they appear in *Nine Thousand Straws: Teaching Thinking through Open-Inquiry Learning* (Knodt, 2008), and document word-by-word what children have shared in the lab setting.

With the guiding design-research question set forth (*Will hands-on open-inquiry build children's ability to think?*), not only does such a program approach inspire curiosity and innovation, it becomes a joyful time and place to share with children. Indeed, open-inquiry sets the pace and spirit for innovation, as one lab student puts it: "Your mind opens up and you want to do all these different things!"

PUTTING COLLABORATIVE INNOVATION TO THE TEST

We had to use teamwork to get the marble to ring the bell. —Lab student

Testing out their creative social risk-taking and problem solving skills, a group of seven fifth-graders has chosen to work with "People and Pipes." This open-inquiry lab challenge is a known stumper: *how will you as a group get a golf ball to run through each PVC pipe (with no hands or pipes touching) and have it land in the metal bucket?* After ten minutes of exploring various possibilities, the group's initial excitement for the challenge has worn thin, with each individual displaying frustration in unique ways. Some twirl the pipes, others use them as horns, while another pair of students stands limp and rolls their eyes. With teamwork dissolving almost on schedule, the instructor smiles a bit to herself as she steps into the scene and is presented with predictable comments from students, such as, *No one is listening to each other! This is a disaster! Lisa is taking over!* And of course, *I'm bored.*

I learned that life is not just handed to you—you've got to figure it out. —Lab student.

> "Children's innovative thinking sets sail when the natural inquisitiveness they bring to the learning table is inspired, affirmed, and cultivated."

The group has agreed on one thing: *This project is impossible!* And with that as a starting point, the lab instructor begins a line of questioning that stimulates an inquiry conversation. *OK, so what is the deal here? How is this challenge impossible? What might help your process of coming up with ideas and thinking together? I'll be back in a bit to see what you come up with!*

It could take a couple more visits from the instructor before the group catches its own stride and works together with the needed thinking verve. Then, further into the session a familiar declaration sounds out in the lab: *We did it! Come see.* To the instructor's surprise (and with his or her renewed confirmation to restrain from leading the problem solving) the group discovers an entirely *new* and remarkably unexpected solution, delighting everyone:

> . . . *Well, we put the tubes against the outer wall so that it wouldn't be as hard. That way they were all at the same level. That made it a lot easier, we just made a few adjustments and it worked!*—Lab student.

Ask About and Explore

With a class of thirty children in the lab, there will likely be ten or more such student-led challenges in action. Along with People and Pipes, a couple of children might examine and classify animal foot prints, while another pair explores mirror symmetry or builds a Rube Goldberg-like chain reaction machine. At a computer station, a few children might be at work with clay and digital images creating a "Claymation" animation. A child may be working alone to build a complex structure with blocks as another child nearby identifies the bones of an upright life-sized human skeleton. Another group of three or more students could be creating a Rebus Story, as another group balances the arms of a mobile with carefully weighed out items, while yet another group designs an "exactly"-twenty-second marble ramp.

Lab Time in a Nutshell

The mission of the described open-inquiry lab is to engage innovative thinking in students by opening up, extending, and guiding the inquisitive energy that children naturally bring to the learning table. At the start, students gather in a circle with a lead instructor and perhaps one or two other parents or teachers. The group spends five to ten minutes discussing a thinking-centered concept that is of interest to all, thus also highlighting the multi-generational learning aspect of the program. As declared by an open-inquiry lab student, which likely all ages can relate to or be inspired by, *"Challenges wake your brain up!"* (Knodt, 2008).

Perhaps the conversation today is about the nature of frustration and finding ways to persevere when things get tough. Such a "focus theme" conversation establishes a thinking-centered objective and dialog possibility to keep in mind (cuing up visiting educators) as lab time unfolds.

At the end of the circle conversation, children make choices about their day's inquiry, with the lead instructor asking: *Where might you find yourself dealing with some frustration today? Where do you want to work?* Children's choices fuel the open-inquiry lab. Engaging the energy and experiences children establish (or fail to establish) with their lab time choices, educators move about the space and interact and instruct by taking on the role of fellow discoverers, conversation builders, and questioning coaches. Whatever the turn-on-the-heel teaching opportunities develop at lab time, the empowering affirmation that starts the action is: *I believe in your choice, I will follow you there.* Essentially, what results is an *apprentice-like* pedagogy. Educators employ the project as a medium through which tools and skills of the thinking trade are guided, discovered, and put into concrete practice.

After a forty-five-minute to an hour lab-time session, the children (even if reluctantly) organize, put their projects back in their places, re-group in the circle, and are asked by the lead instructor: *Let's hear some examples of what was frustrating*

> "The mission of the described open-inquiry lab is to engage innovative thinking in students by opening up, extending, and guiding the inquisitive energy that children naturally bring to the learning table."

> Children *welcome* the opportunity to offer their input and be part of the thinking conversation, especially by using lab inquiries as concrete evidence of their own thinking in action.

with your inquiries today. How did you manage your frustration? How does managing frustration help the process of thinking? What are ways to help a group manage frustration? As the lab session closes for the day, the class is handed a writing-prompt for journal or essay projects to take back to their classrooms: *Describe your inquiry project today. Did you find yourself feeling frustrated? How? What helped you work with your frustration? How could you use those same coping skills when you get frustrated with other challenges?*

Group conversations and students' written reflections confirm the pedagogy in action, as with the student who comments, "I learned not to walk away" (Knodt, 2008). Other comments often surprise educators, who may find themselves learning from the children. For example, there is a lesson in the following comment by a lab student: *"You learn that frustration can be good for thinking"* (Knodt, 2008).

THINKING DISPOSITIONS

Children *welcome* the opportunity to offer their input and be part of the thinking conversation, especially by using lab inquiries as concrete evidence of their own thinking in action. The following is a documented sampling of what open-inquiry lab students had to say when asked: *What does it take to find and meet a challenge?* The answers vary:

- You need to want to do it.
- Patience.
- On some things, you need teamwork.
- Strategy.
- Use what you know.
- Look at the possibilities and . . . the consequences . . . .
- Keep doing it.
- Confidence. Believe you can do it.
- Think ahead. Plan.
- Have a good attitude—don't get grumpy.
- Have the ability to switch ideas.
- Relax.
- Imagine it done.
- Don't be afraid to wonder.

(Knodt, 2008, p. 23)

Thinking About Thinking Dispositions

"Dispositions shape our lives. They are the proclivities that lead us in one direction rather than another within the freedom of action that we have. A thinking disposition is simply a disposition about thinking."

(Perkins 1995, p. 275)

Essentially, what children have identified in the above listing and through their hands-on explorations are *thinking dispositions* or *habits of mind*. An open-inquiry learning community leads children to isolate, understand, and get into practice productive and positive life-long thinking dispositions.

A list of guiding habits of mind can be generated by the teacher-librarian simply by asking children, what are effective habits of mind? Then together they refine a list throughout the year as their work with inquiry unfolds. Another way to create a list is to engage the faculty of the school and arrive at a consensus of ten or fifteen target dispositions. An additional way for the teacher-librarian to establish a list of habits of mind could involve elaborating on the disposition-centered American Association of School Librarian's *Standards* (2007).

The following list of thinking dispositions was developed through lab observations and listening to what children had to say during lab time; through work developed and still underway at Project Zero of Harvard University; as well as from Arthur Costa and Bena Kallick's series on *Discovering and Exploring Habits of Mind* (2006):

- Be Adventurous and Open-Minded
- Wonder, Explore, and Ask Questions
- Contribute Positively to the Group and Inspire Teamwork
- Imagine Possibilities and Outcomes
- Set Goals and Make Plans
- Think Independently
- Use What You Know, Transfer Learning
- Step Back and Look at the Whole Picture
- Strive to Be Accurate and Precise
- Look Carefully
- Listen Actively
- Support Ideas with Reasons Why
- Persevere
- Communicate Clearly
- Understand Others

(Tishman, Jay, & Perkins, 1992; Costa & Kallick, 2000; Knodt, 2008).

With such a line-up, the teacher-librarian frames up a focus on habits of mind that children will use both at school and at home. They may identify and broadcast a monthly or bimonthly habit of mind on the daily news, in school newspapers, or on bulletin boards. This way, not only are thinking dispositions seen in action during inquiry lab time, they become "circulated" throughout all curriculum and children's life experiences. The art specialist might ask, *So our habit of mind this month is imagining possibilities and outcomes. How are you putting a picture in your mind* and visualizing it as you plan your collage in the art room today? The grade-level teacher might ask during a social studies unit, *Class, how do you think Thomas Jefferson engaged the thinking disposition to be adventurous and open-minded?* A parent might ask their family, *How are we communicating clearly as we discuss which movie we would like to see this weekend?*

Again, confirmation of students thinking about thinking unfolds when time is offered for such reflection, as presented by this first grade lab student, "*I need to work on Think First [Set Goals and Make Plans]—because usually I'm thinking so fast I just put something there and the whole thing messes up*" (Knodt, 2008).

As various classes come into the library for their open-inquiry sessions, the teacher-librarian can also customize the lab time by referring to a habit of mind listing. If asked, the grade-level classroom teacher will likely have a thinking disposition or two in mind that would benefit their class through hands-on practice. During this collaboration, the teacher-librarian generates a focus theme discussion, and then both instructors co-teach and work the inquiry floor with the particular disposition in mind. During a lab time circle conversation or back in the grade-level classroom, the group might be asked, *By the way everyone, how does the habit of mind to imagine possibilities and outcomes relate to the book reports we are working on right now?*

FINDING FLOW

Flow, a psychological theory developed by Mihaly Csikszentmihalyi (1990), aims to get at the root "optimal experience." When in a state of flow or a "sustained absorption" with an activity, not only will one lose track of time, one will likely experience strong levels of ownership with the experience.

> "*Optimal experience is thus something that we <u>make</u> happen. For a child, it could be placing with trembling fingers the last block on a tower she has built, higher than any she has built so far . . .*" (Csikszentmihalyi, 1990).

Lab projects are therefore selected by their ability to inspire flow: being process-based, relatively simple, and hands-on. By enlisting Howard Gardner's theory of Multiple Intelligences, lab projects collectively engage and layer a "full spectrum" of intelligences (Csikszentmihalyi & Whalen, 1991; Knodt, 1997, 2008). As a key to teaching for innovation, the objective is for children to get a feel for when flow is in action (or not) while their projects are in hand and to learn to engage such satisfying focus again and again.

Children's flow engagements offer the teacher-librarian, as well as visiting teachers and parents, valuable information that either confirms their understanding of the individual child or offers new perspectives. The inquiry program aims to cultivate a child's curiosity into flow activity, as a goal in and of itself, and also as a base for engaging conversations, focus, and practice with thinking dispositions and their related skills. Additionally, new insights and *trust* established by interacting with students as they work in the lab lead to effective alternative curriculum assessments, alternative GT or LD screening, differentiated learning, and, of course, individualized research and literacy opportunities.

TRANSFER OF LEARNING

If an individual is able to *transfer* what they know from one learning or challenge situation to another, they confirm their ability and understanding of the particular skill or content at hand. Even taking a test is a transfer experience. Yet, many educators do not teach directly for transfer.

> *Without transfer—without this connecting of one thing to another—human learning would not have anywhere near the capacity to shape and empower our lives that it does* (Tishman, Perkins, & Jay 1995, p. 156).

If students are to de-compartmentalize their learning and actively employ grade-level curriculum-based skills, contents, and understandings in different settings, teaching methodologies that establish transfer of learning experiences need to be deliberately engaged by educators (Perkins & Salomon, 2001).

A Transfer of Learning Station

The lab's many constructivist and problem-based projects provide rich opportunities to transfer curriculum-based understandings and skills. As lab work unfolds, instructors perceive possible links and guide students toward establishing their own transfer investigations: *I see you have built a circular tower with the strawberry baskets. I'm curious, how many baskets do you think are in your tower? How could you confirm that hunch? What learning could you transfer?*

Here, a second grade lab student gets the idea of transfer: "*When you are doing a [lab] project, you have to think how it's going to look and what you are going to do first. Or, if you are writing an assign-*

Children's flow engagements offer the teacher-librarian, as well as visiting teachers and parents, valuable information that either confirms their understanding of the individual child or offers new perspectives.

> **Ultimately, one looks for the child to recognize the need or opportunity to transfer the productive thinking dispositions learned and put into practice during lab time, to other situations.**

ment you don't just write something and say you're done with it—you have to decide how you are going to write it and how you are going to plan it out." (Knodt, 2008).

The inquiry program also aims to guide children to transfer their lab-time experience *back out* to support other learning and life arenas as the instructor encourages: *That was a complex bridge you constructed today during inquiry lab time! How did you employ the habit of mind to strive to be accurate and precise as you designed and built it? How could striving to be accurate and precise help with the math problem you are working on now?*

Transfer is a key habit of mind to get into concrete practice for innovative thinking—to aid all learning and unfolding challenge opportunities. Ultimately, one looks for the child to recognize the need or opportunity to transfer the productive thinking dispositions learned and put into practice during lab time, to other situations. Success is hearing a lab student say: *". . . I had to plan a strategy [at lab time] with something. When I had to study at home I remembered back and planned a strategy for the study"* (Knodt, 2008).

BUILDING A HANDS-ON THINKING COMMUNITY

Not only do we as educators need to find new and effective ways to *teach* for innovation, collaboration, and contribution, as we instill the habits of mind to set the pace, we also need to walk the walk ourselves. We need to establish open interactions and "build" instructional programs *with* children, grade-level teachers, specialists, parents, and the community. As libraries establish new traditions for engagement and collaboration, especially along the lines of a *Learning Commons* approach (Loertscher, Koechlin, & Zwaan, 2008), a centralized open-inquiry lab program established by and directly in the library offers many rich possibilities: *"They have to help us build it so they will use it"* (p. 4).

The lab's readily available multi-generational and disposition-centered focus themes, as well as its openness and flow, encourages teachers, administrators, community members, parents, grandparents, and even younger siblings (ages three and up) to walk through the door at any time and work with the children and quickly get on the *same thinking page*. As a result, the centralized open-inquiry learning lab is an ideal time and place to establish a *community* of curious minds in which to build a culture of innovative and industrious thinking *together*.

If you work together, you can think together.—Lab student

Although the teacher-librarian will likely be the one to manage the effort that firmly establishes the program and sets the open-inquiry stage, the program grows strong by establishing co-operative interactions with others and through the help of many.

Building it Together

Support from and collaborations with students start with their lab-time engagement of flow and their contributions to thinking-centered conversations. Once this base-dynamic is established, instructors can guide children's involvements further in a variety of ways, including:

- obtaining students' input, ideas, and leg work for the development of new inquiry challenges and lab projects.
- initiating student peer interviews and questioning about lab-time dispositions in action.
- engaging students as teachers in multi-age groups and mentor programs.

Support from and collaborations with grade-level classroom teachers and specialists include:

- engaging them in thinking-centered focus theme conversations and making supportive links to their various learning arenas.
- co-teaching during lab time to set lab-program learning objectives and methods into action and, as well, establish transfer experiences with grade-level or specialist curriculum goals.
- obtaining grade-level teachers' and specialists' expertise and guidance in the development of new inquiry projects, and making project inventory available for "take out" as inquiry centers for other classrooms.

Support from and collaborations with the students' families include:

- parent volunteers as co-teachers or "Room Guides."
- parents, siblings, or grandparents contributing to focus theme conversations and making home and "life" links.
- parents helping with scheduling, volunteer recruitment, project development, and lab maintenance.

Support from and collaborations with the community at large include:

- inviting area corporate and vocational professionals to work as mentors with children and contribute to focus theme conversations.
- gaining from various vocations and professions new project ideas that help with their development.
- becoming linked with local university

and colleges for potential neurological or cognitive research studies (Fischer, 2009) and for student teaching opportunities.

Such interactions build a dynamic open-inquiry program, but also lead to a "client-based" engagement of potentially all the other media center and library resources available. A start up list of such customized possibilities include:

- guiding children to find follow up resources that support their lab-challenges, such as helping a student who just built a Geodesic Dome with straws and paperclips to find more information about the dome through books on architecture, related web sites, or a biography on its inventor, Buckminster Fuller.
- guiding grade-level teachers, area specialists, and parents toward resources that delve deeper into the concepts and theories put into instructional practice in the lab, such as work with thinking dispositions, project-based learning, Multiple Intelligences, flow, teaching for transfer, or the art of questioning.
- highlighting literature links to thinking dispositions that can be used in the library and other classrooms, such as those available through Arthur Costa and Bena Kallick's, Habits of Mind web site, **http://www.habits-of-mind.net/booklists.htm**.
- linking grade-level teachers, area specialists, and parents with resources useful for project and lesson plan ideas that take their cue from various lab-time challenges.

AFFIRMING CURIOUS MINDS

As becomes clear to an observer of an open-inquiry lab in action, curiosity launches new explorations and sets critical and creative minds into gear. Curiosity and the zest for finding and solving challenges it sparks, can be taught.

I wish I lived here.—Lab student

Children feel connected to a time and place that affirms their choices and builds on the curiosity they have. They give back to the enterprise by helping to establish a dynamic environment full of vital learning energy. Students take ownership, not only of their hands-on challenges, but also of the program itself. As children discover and tell the story of what it takes to be innovative and industrious with thinking, they become part of the community that makes a great library tick.

When it comes to preparing young minds for unknown future challenges and opportunities, this is what the individuals in corporations, policy think tanks, research labs, and the world at large are looking to see—openly curious individuals who are comfortable taking risks, eager to discover new quests, seek collaborations, and are geared toward making contributions with the possibilities they uncover. They are looking for what children intuitively seek to engage and offer to us so freely—the spirit to ask and to explore.

REFERENCES

Costa, A. & Kallick, B. (Eds.). (2000). *Discovering and exploring habits of mind.* Alexandria, VA: Association for Supervision and Curriculum Development.

Csikszentmihalyi, M. (1990). *Flow: The psychology of optimal experience.* New York: HarperPerennial.

Csikszentmihalyi, M. & Whalen, S. P. (1991). *Putting flow theory into educational practice: The key school's flow activities room.* Report to the Benton Center for Curriculum and Instruction. Chicago, IL: University of Chicago.

Fischer, K. W. (2009). "Mind, brain, and education: Building a scientific groundwork for learning and teaching." *Mind, Brain, and Education,* (3) 2-15. Retrieved May 11, 2009, from **http://www.gse.harvard.edu/~ddl/publication.htm**.

Gardner, H. (1993). *Multiple intelligences: The theory in practice.* New York: BasicBooks.

Loertscher, D. V., Koechlin, C., & Zwaan, S. (2008). *The new learning commons: Where learners win!* Salt Lake City, UT: Hi Willow Research & Publishing.

> "Transfer is a key habit of mind to get into concrete practice for innovative thinking - to aid all learning and unfolding challenge opportunities."

Knodt, J. (1997). "A think tank cultivates kids." *Educational Leadership* 55 (1), 35-37.

Knodt, J. (2008). *Nine thousand straws: Teaching thinking through open inquiry learning*. Westport, CT: Teacher Ideas Press.

Perkins, D.N. (1995). *Outsmarting IQ: The emerging science of learnable intelligence*. New York: The Free Press.

Perkins, D.N. & Salomon, G. (2001). "Teaching for transfer." Costa, A., (Ed.). (2001). *Developing minds: A resource book for teaching thinking* (Rev. Ed., Vol. 1). Alexandria, VA: Association for Supervision and Curriculum Development.

Robinson, K. (2006). "Do schools kill creativity?" TED, Ideas worth spreading. (Video). Retrieved May 10, 2008, from http://www.ted.com/index.php/talks/ken_robinson_says_kill_creativity.html.

Tishman, S. & Andrade, A. *Thinking dispositions: A review of current theories, practices, and issues*. (Paper). Retrieved January 22, 2008, from http://learnweb.harvard.edu/alps/thinking/docs/Dispositions.pdf.

Tishman, S., Perkins, D. N., & Jay, E. (1995). *The thinking classroom. Learning and teaching in a culture of thinking*. Boston: Allyn and Bacon.

Jean Sausele Knodt, an artist and educator, directed and designed an open-inquiry learning lab in a Fairfax County, Northern Virginia public school for nine years that served over 700 regularly attending K-6 students. Knodt currently gives presentations and workshops on open-inquiry, teaches Fine Arts, and is working on designing inquiry practices for the college/university level. Knodt is author of *Nine Thousand Straws: Teaching Thinking through Open-Inquiry Learning*, reviewed in the June, 2009 issue of *TL*. She may be reached at *inspired.minds@rcn.com*.

POSTER REFERENCES:

Costa, A. & Kallick, B. (Eds.). (2000). *Discovering and exploring habits of mind*. Alexandria, VA: Association for Supervision and Curriculum Development.

Csikszentmihalyi, M. (1990). *Flow: the psychology of optimal experience*. New York: HarperPerennial.

Csikszentmihalyi, M., & Whalen, S. P. (1991). *Putting Flow Theory into educational practice: The key school's flow activities room*. Report to the Benton Center for Curriculum and Instruction. University of Chicago.

Gardner, H. (1993). *Multiple intelligences: The theory in practice*. New York: Basic Books.

Knodt, J. (1997). "A think tank cultivates kids." *Educational Leadership* 55 (1): 35-37.

Knodt, J. S. (2008). *Nine thousand straws: Teaching thinking through open-inquiry learning*. Westport, CT: Teacher Ideas Press.

Perkins, D.N. (1995). *Outsmarting IQ: The emerging science of learnable intelligence*. New York: The Fee Press.

Perkins, D.N., & Salomon, G. (2001). "Teaching for transfer." Costa, A. (Ed.). (2001). *Developing minds: A resource book for teaching thinking*. Alexandria, VA: Association for Supervision and Curriculum Development.

Tishman, S., Jay, E., & Perkins, D. N. (1995). *The thinking classroom. Learning and teaching in a culture of thinking*. Boston: Allyn and Bacon.

Poster. Copyright © 2008 by Jean Sausele Knodt. Libraries Unlimited/Teacher Ideas Press. Reproduced with permission of ABC-CLIO, LLC.

Feature articles in *TL* are blind-refereed by members of the advisory board. This article was submitted May 2009 and accepted July 2009.

FEATURE ARTICLE

SCHOOL LIBRARY 2.0: FROM THE FIELD

information literate? just turn the children loose!

EDITOR'S NOTE: IT IS WIDELY ASSUMED THAT CHILDREN AND TEENS MUST BE FURNISHED WITH AN INFORMATION LITERACY MODEL THAT GUIDES THEIR STEP-BY-STEP PROGRESS THROUGH A RESEARCH ASSIGNMENT. IN ADDITION, THE FURNISHED MODEL IS TAUGHT OVER AND OVER AS LEARNERS PROGRESS THROUGH THE GRADES. RARELY HAVE WE SEEN IN THE LITERATURE ENCOURAGEMENT TO HAVE LEARNERS REFLECT ON THAT MODEL. JOY MOUNTER NOT ONLY HAS FIRST-GRADERS REFLECT ON A MODEL THEY ARE USING FOR RESEARCH, BUT AFTER REPEATED EXPOSURE, THE LEARNERS THEMSELVES BECOME RESTRICTED BY THE MODEL AND INVENT THEIR OWN. IT IS THE PROCESS OF TAKING COMMAND OF ONE'S OWN LEARNING THAT IS BOTH REMARKABLE AND AN ESSENTIAL ELEMENT OF THE LITERACY PROGRAM OF THE LEARNING COMMONS.

their elementary school years. For this notebook and in an enlarged version, I used the TASC (Thinking Actively in a Social Context) model of Belle Wallace (author and past president of the United Kingdom's National Association for Able Children in Education) and built a model children could begin using as pictured in Figure 1. The original model has the learner ask the following questions when faced with a research assignment:

1. What do I know about this?
2. What is the task?
3. How many ideas can I think of?
4. What is the best idea?
5. Let's do it!
6. How well did I do?
7. Let's tell someone!
8. What have I learned?

What happened over time as the students recorded in their journals was amazing. They began to reflect as they grew and developed, understanding

My journey of hope and change began with the arrival of "Excellence and Enjoyment." I remember reading the first page and introduction by Charles Clarke that said, "There will be different ways. Children learn better when they are excited and engaged. . . . Different schools go about this in different ways" (DFES, 2003, p. 3).

For the first time since the rigidity of the literacy hour, I felt we had hope to really make changes that would matter and have an effect on our students by developing creativity and flexible ways of thinking and learning about learning itself. To stimulate metacognition for every child in my class, I developed a spiral notebook in which they recorded their personal learning journeys across

joy mounter

FIGURE 1

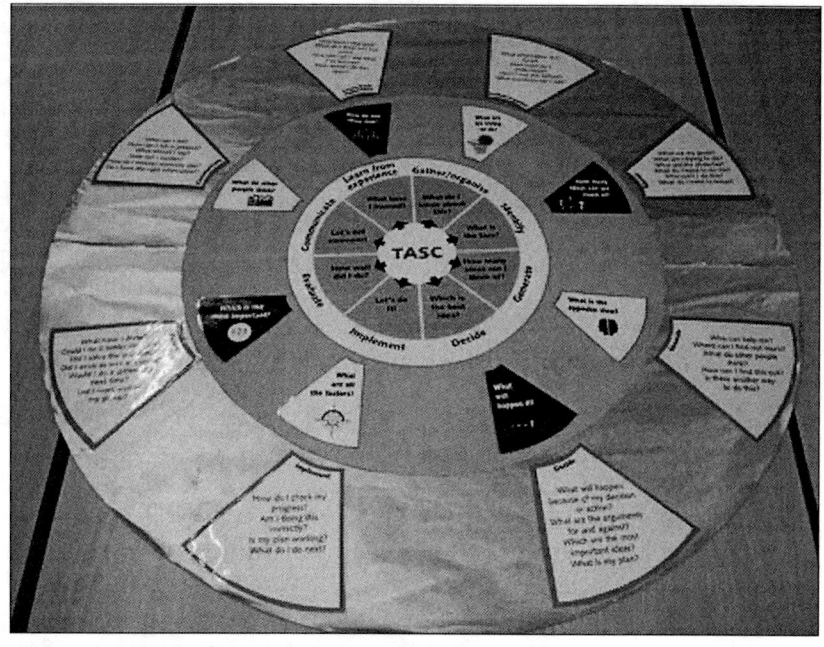

www.webquestuk.org.uk/TASC%20WHEEL/Wheel.htm

SCHOOL LIBRARY 2.0: FROM THE FIELD

themselves—the things that make them tick, their worries and strengths, as well as the quirks that make them an individual. They began to form their own learning values and articulated them to others.

Art Costa highlights that "all human beings have the capacity to generate novel, clever or ingenious products, solutions, and techniques—if that capacity is developed" (Costa & Kallick, 2000, p. 32). That is what I wanted the children to develop, to have the opportunities to work creatively on any task, thinking outside of the box, with freedom and risk taking; to see learning as being flexible and fluid and that requires different skills and responses; to think beyond their immediate learning and begin to generalize and create their own theories of learning. For me, personalizing learning means enabling a child to react to any learning situation with an understanding of self and the ability to empathize and evaluate, working with the learning skills of others around them.

But this has to be in the context of a learning environment and a creative curriculum where the children are involved in developing their own educational theories. I wanted to develop a format for the children to explore learning, including theories of others, and use this as a platform to create their own knowledge and values.

The excitement is sharing the journey with my class. They look at the world through very different eyes than I. We talked about the research I was conducting on learning. They were surprised and challenged me immediately, asking if I was writing about learning, didn't I need their help? From the tone of the comment, it was clear the children could not comprehend that I, or anyone, could write about learning without asking for their help. It made me take a step back and look at learning in my classroom from a different perspective. Remember: These children are just aged 6, 7, and 8 (we are a mixed Year 2/3 class). Their ideas were thought provoking and challenged my thinking, which helped me see as a learner through their eyes.

We quickly made a large wheel (Figure 1), which we kept on the classroom wall. Initially it was used a lot, but soon the flow of the segments became integral

FIGURE 2

The discussion of the children talking about their thinking about the TASC Wheel can be viewed on YouTube at www.youtube.com/watch?v=hH2-5xexbAQ.

and embedded with the students. We started with a topic-focused inquiry week using the wheel to plan and implement throughout the week. This enabled us to see all of the segments of the wheel working and supporting each part in a small time frame. It enabled everyone to be involved, to clearly understand the steps we were taking, but didn't contain our creative learning. Rather, it allowed us to fly but gave us the vocabulary and ideas as prompts.

From that moment, the children and I saw the value of the wheel and loved using it for its simplicity but also for the layers of thinking it encouraged and challenged us to use. Soon the wheel became the framework for all our work. Even if it was not explicitly talked about, it had become so embedded that we talked and planned using the format confidently. Other uses were quickly found for short planning of topics, which we called topic plans, in science or any area of the curriculum. Using the wheel as a whole or in smaller parts focused our evaluation, the children felt more confident knowing it was on the wall, and they actually saw it as a resource if they were "stuck." They would look at the flow of the segments and the focus questions on the inner wheel to help them. I asked the children to record the ways they could help themselves if they were stuck, they thought of the wheel first.

It seems almost strange looking back at the beginning of our journey. The TASC Wheel stimulated the children's interest in the adult's writings about learning. We critically explored the writings of Belle Wallace and even wrote to her, challenging some of her ideas. While the children were confident exploring learning skills and using the TASC Wheel, they were dissatisfied as well. Child P described the wheel as too flat, too two-dimensional; whereas his thinking spirals around, flows over the edge of the circle and up through the middle of the wheel, and explodes, sometimes showering others with sparks from his learning. Other children described the wheel almost as a cushion, with no outer edge—more of a curve before spiraling our thinking up through the middle. They began to recognize they would never be in the same place again as learners. Figure 2 is the result of the children's discussion of thinking and learning, and Figure 3 reports on their progress. Figure 4 is the wheel the children developed from their experiences and reflections as learners.

74

SCHOOL LIBRARY 2.0: FROM THE FIELD

FIGURE 3

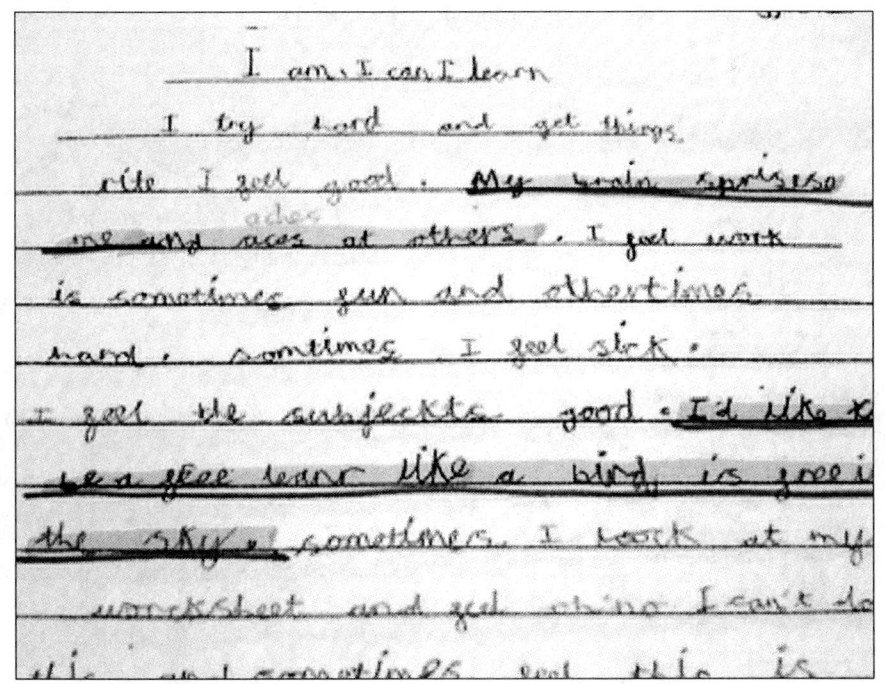

[handwritten text: I am, I can, I learn / I try hard and get things rite I feel good. My brain surprises me and aces at others. I feel work is sometimes fun and othertimes hard. sometimes I feel sick. I feel the subjects good. I'd like to be a free learnr like a bird, in the sky. sometimes I look at my worksheet and feel oh no I can't do this...]

Often we got caught at the end of the wheel, and it was easy to share our learning on a topic through an assembly. We thought about the facts we had learned, that the wheel should encourage us to self-reflect, and evaluate skills we have used and need to develop within the topic. It is the knowing of "self" that moves our thinking forward and the emotional aspects of self that reflect in learning.

For us, the last two segments of the wheel were most important, the ones we learned most from and with. And perhaps they should be the start of all reflections leading to learning and not seen as the end. A slight turn of the wheel, and they appear at the top and the beginning and not the end.

SOMETHING TO SAY, SHARE, AND LEARN

During one of our conversations as a class, we again returned to Belle Wallace's TASC Wheel. The children were curious to know whether lots of schools used the TASC Wheel. I took the opportunity to introduce the vocabulary of having a learning theory. This was a difficult concept for the children to comprehend, and we struggled together for a while. From the discussion came the idea that the children wanted to have a learning theory of their own. As experts, they felt people should listen to them and not adults, as it is children who are the learners all day and for years as they grow.

The talk about theories had awakened a keen need to begin planning and articulating their ideas to form a learning theory of our own. Following the idea that TASC meant something when you looked at each letter, the children talked in pairs for a special word of their own to summarize the learning theory. I was amazed as very quickly Child A suggested the word *QUIFF*. The children liked the sound of the word and began thinking what the individual letters could stand for, just like TASC. They didn't have to argue or even debate ideas; they quickly agreed, and all ideas seemed to come from the group, almost as a collective mind.

• Questions we all have to ask to learn
• Understand—making sense of things around us and ourselves, which is harder
• I am important
• Feelings so important as a learner
• Focus to be able to concentrate and persevere

QUIFF, "I" as in "I am important" is in the center, just as we are the center of our learning and self. "I" is surrounded by our understanding of "things" and ourselves, feelings that often control our learning, and focus—applying ourselves as learners.

The class then decided that as TASC is represented by a circle, they needed a visual image for QUIFF. I quickly handed out paper for them to draw their ideas, and those thoughts turned to the shape for QUIFF. The resulting pictures were all so different and very thoughtful. Figure 4 shows Child A's picture. She used a triangle with "I" at the point, represented by an eye—the most important point: an eye to the world

FIGURE 4

and into ourselves. Question marks are at the bottom, the start and widest part of the shape. Focus is like an egg floating in between our questions and feelings that control our thoughts, our learning, and us.

Kellett (2005) highlights the opportunities for pupils to engage with a subject in great depth and work with primary, self-generated data. The depth of the children's thinking shocks anyone we share our journey with. Age, knowledge, and skills have often been quoted as barriers to children taking part in action research successfully, but this study will challenge these preconceptions, encouraging the children to critically challenge each other's thinking and funnel down their research questions and test their hypotheses.

Following this session, we recorded our questions and thoughts, defining our ideas behind QUIFF. We began thinking about how children learn best, generating ideas, sharing them, and then discussing and recording those they all agreed on. We learn best when we:

- understand and use our learning skills
- believe in ourselves
- think about ourselves as a learners
- are curious
- are happy and calm

Of course, with each new group, a new journey unfolds. We can use the older children who have been through the process as learning coaches, but we must be open to new ideas, models, drawings, and developments as the new groups travel their own journey. Our journey enabled us to develop our understanding of ourselves as learners and as people. We reflectively challenged ourselves and challenged others. The children and I wonder if adults are ready to listen?

REFERENCES

Costa, A. L., & Kallick, B. (2000). *Discovering and exploring habits of mind*. London: Association for Supervision and Curriculum Development.

DFES. (2003). *Excellence and enjoyment: A strategy for primary schools*. London: DCSF Publication Centre.

Joy Mounter is head teacher of a seven-class primary school on the outskirts of Shepton Mallet in Somerset, England. She may be reached at *joymounter@aol.com*.

FEATURE ARTICLE

Gifted Readers and Libraries: A Natural Fit

The library provides a place where the world can come to the student.

REBECCA HASLAM-ODOARDI

Jenna, a fourth grader in Mrs. Sondberg's class loves the library. She thinks it is the best place in the entire school.

"Mrs. Wilcox [the teacher-librarian] helps me find the books I want to read," Jenna shares. "She knows I have already read most of the books the fifth and sixth graders read. So she always buys new books for me to read that are challenging and exciting!" And that's not all.

When prodded to discuss any other reasons why she likes being in the library, Jenna enthusiastically mentions that she likes to have "book discussions" with Mrs. Wilcox. "What are those?", I ask. "Well" she pauses, thinking, "Mrs. Wilcox understands that to keep me engaged in reading, I need to think deeper about what I read. She helps me do that by asking me questions that I have to think really hard to answer."

"Does she do this with all the students?" I ask.

"Yes, but with me and some of the other higher level readers, she asks us questions that are harder."

"How do you like that?" I continue.

"It's the very best part of being in the library. I love the "group think" time with Mrs. Wilcox because I love to read and reading shouldn't just be about hurrying through the book. It should be about taking time to study what the book is really about."

Not all advanced readers are as lucky as Jenna, who has a teacher skilled in challenging her reading abilities. Research has shown that the failure to provide appropriate reading instruction for gifted readers resulted in a decline in positive attitude toward reading, especially in the higher grades (Swiatek & Lupkowski-Shoplik, 2000). Gifted students need differentiated curriculum that is tied to their interests and abilities in order to be well-functioning in the school environment. Schools that do not attend to the needs of their students run the risk of the students becoming apathetic, disengaged, and perhaps even dropping out of school (Archambault, Westberg, Brown, Hallmark, Emmons, & Zang, 1993; Colangelo et al., 2004; Renzulli & Park, 2002; Reis et al., 2004).

Data from the Utah Advanced Readers At Risk (ARAR) program show that providing instruction to teachers about how to differentiate for reading instruction and the specific content to teach during that reading instruction made a difference both in the lives of the students as well as in the teaching abilities of the teachers (Hunsaker, Bartlett, & Nielsen, 2009). Advanced readers had the chance to read books that interested them and that were at an appropriate level of challenge for them, and teachers learned the strategies to use with these students so they continued to grow in the area of reading.

THE ADVANCED READERS AT RISK PROJECT

Jenna's teacher was involved in a Jacob K. Javits Gifted and Talented Education Federal Grant. The purpose of the Advanced Readers At Risk: Rescuing an Underserved Population (Hunsaker, S.L., Odoardi, R.H., et. al, 2003) grant awarded to Davis County School District from the United States Department of Education (2003-2007), was to provide appropriately challenging reading opportunities to those students who read significantly above grade level. Teachers involved in the project were recruited from Utah school districts. They were divided into two cohorts: Cohort I participated in years one, two, and three, and Cohort II participated in years two and three. Cohort I began with 30 teachers who taught fourth, fifth, and sixth grades. The 31 Cohort II teachers also taught fourth, fifth, and sixth grades. The project included teachers from both Title I and non-Title I schools.

WHO ARE THE HIGH-ABILITY LEARNERS?

As the program began, the teachers naturally wondered who these students were. Were we talking about the child who read voraciously and did not want to do anything else? Were we talking about those students who tested well? Were we talking about the students who said they "loved" reading? What we found was that all of the above could be said about advanced readers. They may be the students who come into the library, sit down, and read voraciously. They may be the students who wander around the library for the entire library time trying to choose a book to read. And, they may also be the students who continually ask questions about reading, about books, and about the library itself that seem more advanced than those asked by the other students in the class.

We know that gifted readers are early and voracious readers, have advanced vocabularies, and perform better on reading assessments than their age level peers (Vacca, Vacca, & Gove, 1991). In addition they use words easily, accurately, and creatively in new and innovative contexts and they perceive relationships between and among characters and grasp complex ideas (Collins & Kortner, 1995; Halstead, 1990). There is also strong evidence that they may not benefit from conventional reading instruction (Catron & Wingenbach, 1986; Dooley, 1993; Levande, 1999). They need to be able to explore books that answer their many questions, books that tease and invite introspection, and they need to be encouraged to read what they love in order to continue to read. The teacher-librarian, central to all of these purposes, was an important partner in the Advanced Readers at Risk project.

THE FOUR COMPONENTS OF THE ARAR READING PROGRAM

The four components of the model are shown in Figure 1. *Learning to Read* and *Reading to Learn* have to do with the acquisition of advanced academic skills and *Reading to Serve* and *Reading for Leisure* describe what gifted readers do as they interact with what they read. Two of the components of the model—learning to read and reading to learn—are designated as "academic" reading because they are primarily applied for purposes of schooling. The other two components—reading for leisure and reading to serve—are considered "active" reading because they are used actively for personal purposes. There is also

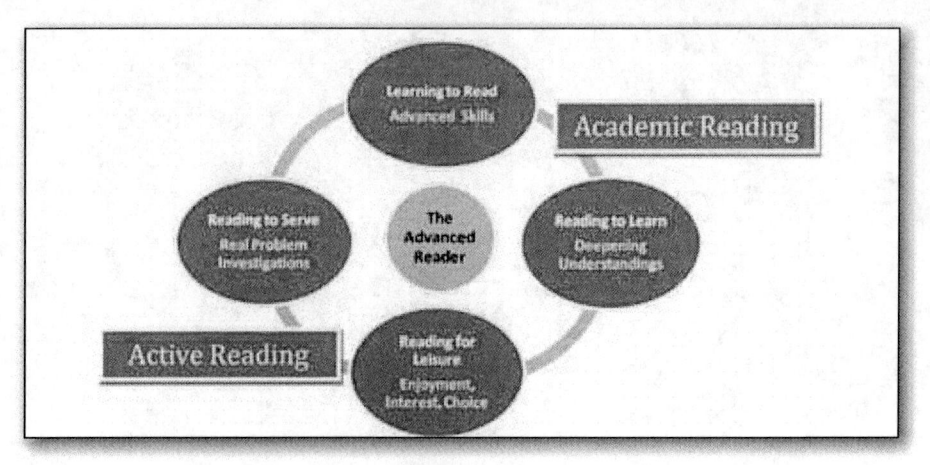

Figure 1. The Advanced Readers at Risk Program Explained.

a relationship among the components that exists along the diagonals where learning to read and reading for leisure are considered intrinsic because reading is for reading's sake, whereas reading to learn and reading to serve are more instrumental because reading is a skill for purposes other than reading itself (Hunsaker, 2002).

SELECTING BOOKS FOR ADVANCED READERS

How do we select books that have advanced vocabulary and content and yet are still appropriate for the student's reading level? Jenna, at age ten, was ready to read very complex literature, yet the teacher and teacher-librarian had to work in concert to be sure the literature she chose was appropriate. What then are the ideas teachers and teacher-librarians need to consider when helping advanced readers find books that are appropriate for them?

First, books for advanced readers should have strong characters gifted readers can relate to and/or characters they can admire and emulate. In the book, *The Watsons Go to Birmingham-1963* by Christopher Paul Curtis (1995), a Northern black family travels to Birmingham, AL during the height of the Civil Rights movement and witness one of the darkest events in American History. The book is an excellent example of courage, tenacity, and determination. The characters are strong and their virtues are those we would want students like Jenna to emulate. As an assignment, students could write biographical sketches of these characters or compare and contrast what they see in them with their own life in today's world.

In the newly released, *A Season of Gifts* by Richard Peck (2009), the character of Grandma Dowell is intriguing. Her gutsy, no nonsense personality, fearless sense of right and wrong, along with her subtle sense of humor are fascinating. This book, the third in the series of Grandma Dowell books, is full of unique phrases and historical references that set the stage for excellent student- or teacher-led discussions. Questions about phrases such as "gray in the gloaming," "I'm techy as a bull in fly time," or "the silence fell upon the listening town" helps students explore the subtleties and nuances of language.

Second, when choosing books for advanced readers, the language used in the book needs to enrich the text as it challenges, stimulates, and stretches the reader. Vocabulary that is new and challenging asks the advanced reader like Jenna to think, to consider the context of the sentence, and to discover new words. In Michael Clay Thompson's *100 Classic Words* (2006) he suggests exposing students to words like traverse, repose, lurid, superfluous, sagacity, tremulous, wan, indolent, maxim gives them the background information they need to read more sophisticated texts. Jenna needs to be able to talk about the meanings of these words with her teacher and with her peers. She needs to use them in her writing and she needs to be encouraged to broaden her own vocabulary. The following from the classic book, *Call of the Wild* by Jack London (1903) is an example of language that is rich and vivid.

Third, books that are appropriate for advanced readers need to involve plot that is complex. It is often the complexity of a book that makes it most challenging for the advanced reader. Important questions to ask would be:

• Does the story present a multiplicity of ideas or concepts?

• Does it warrant deep thinking about a topic that is interesting and important?

• Is the story predictable or does it lead the reader to wonder, to suppose, and to imagine?

One of the most intriguing things about the book *Tunnels* by Roderick Gordon and Brian Williams (2007) is that the reader always encounters the unexpected. The twists and turns in the plot keep the reader guessing, wondering, and yes, reading! Exploring these twists and turns, diagramming the rising and falling action, discussing the conflicts experienced by the characters in the story, exploring how the author brings resolution to these conflicts, and also noting the techniques used by the author to develop the climax are ways to not only improve gifted readers skills in reading but also help them understand and improve their writing skills.

Fourth, books for advanced readers need to employ a variety of literary devices. Onomatopoeia, alliteration, assonance, flashback, forecasting, satire, irony, and hyperbole are examples teachers can use with advanced readers. Because of their cognitive abilities, these students need these strategies so they will think more deeply about the connections they can make to the text. Exposing gifted readers to poetry and poetry anthologies are excellent ways to introduce them to literary devices.

In the rerelease of Robert Frost's poetry collection titled, *You Come Too: Favorite Poems for Young Readers* (2002), students have the opportunity to explore the literary devices found in his work. As gifted students read this book, they could be asked to:

• Identify the literary devices in the poetry;

• Create a new poem that expresses one of the key ideas in Robert Frost's poem; and

• Analyze the poetic form in one of his works and write a new poem using that form.

Fifth, advanced readers need to be encouraged to read all kinds of literature—fiction and nonfiction. They need to explore the world through books about geography, science, and history. They need to be encouraged to question, to wonder, and to find the answers to their questions through the literature they read. Internet resources, blogs, and webcasts may prove to be the best resources for advanced content for these students; however, libraries also need to include books and materials that are more advanced. The award winning nonfiction book, *Freedom Riders: On the Front Lines of the Civil Rights Movement* by Ann Bausum (2005) is the story of two young men, one white and one black who boarded a bus in the south to aid the cause of civil rights. The challenges they encounter, the attitudes they find in the people of the south, and also the heroism they display in a time of great political unrest not only exemplify courage and commitment but also provide gifted readers with opportunities to connect both fiction and nonfiction books to the history they are studying in school. Students can explore the theme of "conflict" by both analyzing and comparing the fictional book *The Watsons Go To Birmingham"* with this nonfiction account.

Activities might include:

• Creating a multimedia presentation about the major events during the Civil Rights movement using examples from both books.

• Exploring the laws and court actions that we have today as a result of the Civil Rights Movement.

• Discovering how the political unrest of the 1960's led to today's election of an African American for President of the United States.

While activity differentiation with the same learning objective may require a teacher-librarian to provide appropriate information to the student's needs (in this case potentially beyond grade level), advanced readers need to have this support to freely explore their interests. They also need to be given credit for the curricular concepts they already know and be given opportunities to focus on in-depth study projects like those mentioned before.

Sixth, advanced readers need opportunities to read about events, situations, and circumstances going on in their world that interest them. Because of their ability to read material that is more advanced, they may have questions about local, state, or world situations that other students either are not yet interested in or do not understand. The library provides a place where the world can come to the student. Reading materials that support their desire to know and understand are important for these students. As teachers and teacher-librarians consider books for the advanced readers in their classroom, a checklist like the one in Figure 2 might prove helpful.

HELPING ADVANCED READERS CONTINUE TO LOVE FOR READING

In the Advanced Readers at Risk project, we found that advanced readers do not always love to read.

Some strategies we have used with advanced readers either hold them back or do not help them to learn new things. Advanced readers need to be encouraged to read material that may appear difficult but will always expose them to something new. This also means they may need adjustments in when reading assignments are due so they can have the time they need to read more difficult selections.

For instance, accounting models for recreational reading—programs in which books, pages, minutes, or points are counted, especially where such programs limit students to specific reading lists, do not allow advanced students the opportunity to select literature that interests them and is appropriately matched to their ability.

Jenna's teacher used discussions, student interviews, and her own information on how Jenna responded to advanced questions to assess whether or not Jenna and the other advanced students were acquiring new reading skills. By using strategies that helped the students acquire new reading skills, Jenna's teacher was able to interrupt the declines in interest advanced readers typically show toward reading. By letting students have more choice in what they read, by eliminating accounting models, and, instead, finding ways to ignite the passion for reading, teachers and teacher-librarians in the ARAR project

	Things to Consider When Selecting Books for Advanced Readers
√	Strong characters advanced readers can relate to and/or characters they can admire and emulate.
√	Language that enriches the text and challenges, stimulates, and stretches the reader.
	Complexity in plot structure.
	Variety of literary devices.
	Selected from a broad range of genres.
	Interesting to the student.

Figure 2. Checklist to help in book selection

were able to develop and maintain students' love of reading.

The teacher-librarian partnered with Jenna's teacher to focus on the joy of reading. She read aloud to small groups of students who were grouped according to their need and ability. The teacher-librarian worked collaboratively with the teacher to pique their interest with various books by giving book talks, by encouraging questions about the books, and by guiding the students to choose appropriate books for successful, challenging reading. The book discussions were always on various topics of interest to the students and the higher-level questions like those shown in Figure 3, were used to encourage the advanced readers to think at a deeper level.

LIBRARIES AS HOME FOR GIFTED READERS

The library became Jenna's favorite place because her teacher-librarian encouraged and challenged her.

Through the ARAR project it became clear teachers, gifted and talented specialists, and teacher-librarians need to work together as a learning community to provide appropriate instruction for the advanced readers in the classroom. Together they need to talk openly about the materials, books, and information needed to meet the variety of ability levels in the classroom. They need to discuss how the time spent in the library could extend and enrich the learning for all students, but they also need to work collaboratively to guide and direct learning for the advanced reading student so that, like Jenna, advanced readers know the library as a friendly, open, engaging place where opportunities are limitless.

BOOKS MENTIONED

Bausum, A. (2005). *Freedom riders: On the front lines of the Civil Rights Movement.* Des Moines, IA: National Geographic Children's Books.

Curtis, C.P. (1995). *The Watson's go to Birmingham-1963.* New York: Random House.

Frost, R. (1959, rerelease 2002). *You come too: Favorite poems for young readers.* New York: Holt.

Gordon, R. & Williams, B. *Tunnels* (2008). New York: Chicken House, Scholastic.

London, J. (2001). *Call of the Wild.* New York: Simon & Schuster.

Peck, R. (2009). *A Season of Gifts.* New York: Dial Books for Young Readers.

REFERENCES

Archambault, F. X., Jr., Westberg, K. L., Brown, S. Hallmark, B.W., Emmons, C. L., & Zhang, W. (1993). *Regular classroom practices with gifted students: Results of a national survey of classroom teachers.* Storrs, CT: The National Research Center on the Gifted and Talented.

Catron, R.M., & Wingenbach, N. (1986). Developing the gifted reader. *Theory into Practice*, 25(2), 134-140.

Bloom's Taxonomy Level	FICTION QUESTIONS
Analysis	What would you infer is the author's purpose for writing the book?
	What ideas in the book validate the author's choice of the book's title?
	Discuss the pros and cons of changing the ending of the book?
Synthesis	What additional facts could you gather to help the author extend the story?
	How does the subject of the book connect to other books you have read?
	What changes would you make to revise the beginning of the book?
Evaluation	What criteria would you use to assess the worth of this book?
	What is the most important concept you think the author was trying to teach the reader?
	Share a choice that you would have had a character make that is different from the one the author chose.

Bloom's Taxonomy Level	NONFICTION QUESTIONS
Analysis	Infer what you believe is the most important information the author wants the reader to know about this topic.
	Survey your friends and family to find out why this topic is important for people to consider.
	What projects can you suggest as a result of your reading about this topic?
Synthesis	Make a presentation to the class about the effect this topic has or could have on them.
	Generate a poster or campaign dealing with some aspect of the topic you read about.
	Make a list of careers or job opportunities you think could result from the topic of your book.
Evaluation	How could you determine whether or not the information presented in the book is true or false?
	What resources or new ideas could the author have included in this book that seems to be left out?
	Prepare a debate to defend your position regarding some part of the book.

Figure 3. Sample Questions/Suggestions for Advanced Readers

17 Don'ts for Gifted Readers By Rebecca Odoardi

1. Don't think they know everything. They still need to learn new reading strategies.

2. Don't be afraid to challenge them to read difficult material.

3. Don't force them to read books that are too easy for them. On the other hand don't discourage them from reading books that are easy for them when 'comfort' reading is the goal.

4. Don't make them count pages or time...doing this may lead to poor attitudes toward reading.

5. Don't use computer reading programs to track their progress or assess them unless you are willing to write a test for more difficult books they may choose to read. This may create a situation where they read an easier book just to accumulate points.

6. Don't be afraid to group for instruction...whole group instruction does not meet everyone's needs.

7. Don't group them with students who are far below their level or make them be the group "teacher" for these students. They aren't teachers!

8. Don't be concerned if they sometimes make mistakes. Gifted readers aren't perfect!

9. Don't forget to involve them in sophisticated dialogue about what they read; use higher level Blooms taxonomy questions with them.

10. Don't always choose books for them to read. Gifted readers like to choose the books they want to read.

11. Don't just "assign" books; give them reasons to read books through book talks and other inviting book activities.

12. Don't feel they always have to defend their reading choices. Let them sometimes read something just because they want to read it.

13. Don't forget to encourage them to find the answers to their questions—to research answers on the Internet, through blogs, and interviews. Teach them how to do sophisticated research at a level they can handle.

14. Don't inhibit their quest for information. Gifted readers often want to find the answers to questions other students aren't mature enough to ask. Carefully guide them to appropriate resources for answers to their questions.

15. Don't forget they need time to discuss their ideas, insights, and perceptions with their *intellectual* peers.

16. Don't make them finish every book they start. Give them the right to decide that a particular book is not good for them.

17. Don't discourage them from rereading a book they love. Even adults sometimes reread favorite books.

Colangelo, N., Assouline, S., & Gross, M. (Eds). (2004). *A nation deceived: How schools hold back America's brightest students.* Iowa City, IA: The University of Iowa.

Collins, N.D., & Aiex, K. N. (1995). Gifted readers and reading instruction. ERIC Digest (Report # ED379637). Retrieved September 9, 2009 from http://www.eric.ed.gov/ERICWebPortal/custom/portlets/recordDetails/detailmini.jsp?_nfpb=true&_&ERICExtSearch_SearchValue_0=ED379637&ERICExtSearch_SearchType_0=no&accno=ED379637.

Dooley, C. (1993). The Challenge: meeting the needs of gifted readers. *The Reading Teacher,* (46)546-551.

Halsted, J.W. (1990). Guiding the gifted reader (Report # ED321486). ERIC Digest E481. Retrieved September 10, 2009 from http://www.eric.ed.gov/ERICWebPortal/custom/portlets/recordDetails/detailmini.jsp?_nfpb=true&_&ERICExtSearch_SearchValue_0=ED321486&ERICExtSearch_SearchType_0=no&accno=ED321486.

Hunsaker, S.L., Bartlett, B., & Nielsen, A. (2009). *Correlates of teacher practices influencing student outcomes in reading instruction for advanced readers.* Manuscript submitted for publication.

Hunsaker, S.L. (2002). Opportunities and challenges for the gifted reader. The world-class reader model. *Gifted and Talented, 6*(1), 16-18.

Levande, D. (1999). Gifted readers and reading instruction. *CAG Communicator,* 30(1), 19-20, 41-42.

Reis, S. M., Gubbins, E.J., Briggs, C., Schreiber, F.R., Richards, S., Jacobs, J., Renzulli, J.S. (2004). Reading instruction for talented readers: Case studies documenting few opportunities for continuous progress. *Gifted Child Quarterly,* 48, 309-338.

Renzulli, J.S., & Park, S. (2000). Gifted dropouts: The who and the why. *Gifted Child Quarterly,* 44, 261-271.

Swiatiek, M.A., & Lupkowski-Shoplik, A. (2000). Gender differences in academic attitudes among gifted elementary education students. *Journal for the Education of the Gifted,* 23, 360-377.

Thompson, Michael C. (2006). Classic words. Retrieved Sept. 30, 2009 from http://www.rfwp.com/downloads.php#4.

Vacca, J., Vacca, R., & Gove, M. (1991). *Reading and learning to read.* New York: HarperCollins.

Rebecca Haslam-Odoardi is an educational consultant in the field of gifted/talented education. She retired recently after eighteen years as Director of Gifted and Talented programs in Davis County, Utah, one of the largest school districts in the state. She has been President of the Utah Association for Gifted Children and serves as Co-Chair of the Legislative Committee for the National Association for Gifted Children. She may be reached at *rodoardi@hotmail.com.*

FEATURE ARTICLE

SCHOOL LIBRARY 2.0: FROM THE FIELD

using the library learning commons to reengage disengaged students and making it a student-friendly place for everyone

EDITOR'S NOTE: OUTSIDERS SETTING UP OFFICE IN THE LIBRARY? ENCROACHMENT ON LIBRARY SPACE? THIS ACCOUNT OF A "HOMELESS" STUDENT SUCCESS SPECIALIST WHO SEES LITTLE CORRELATION BETWEEN HER ROLE WITH AT-RISK KIDS IN THE LIBRARY SHOWS THAT WHEN SHE AND THE TEACHER-LIBRARIAN DECIDE TO MAKE THE BEST OF A JOINT OFFICE FACILITY, MARVELOUS THINGS HAPPEN. OUR TEAM OF SPECIALISTS DISCOVERS THEY HAVE OPPORTUNITIES TO SUCCEED FAR BEYOND WHAT THEY COULD DO ALONE. DID THE TEACHER-LIBRARIAN GAIN ANOTHER PROFESSIONAL IN THE LIBRARY? OR, DID THE STUDENT SUCCESS SPECIALIST GAIN ANOTHER PROFESSIONAL FOR HER PROGRAM?

When a student at Adam Scott CVI (Collegiate and Vocational Institute) is struggling because of emotional, social, or behavioral issues, they do not go to the office—they go to the library learning commons.

When I was in high school in the 1980s, I could not wait for Grade 13. At my high school, the university-bound graduating students were allowed access to the balcony in the library for "quiet study." The balcony, furnished with desks, chairs, couches, and tables, was off limits to all other grades. Younger students sometimes tried to sneak in but faced immediate eviction.

The library was commandeered by a woman who maintained strict order at all times. Talking was taboo, and even spirited whispering was discouraged. The library was for serious-minded academic students who worked silently and followed the many library rules. The library catered to a certain kind of student, and if you didn't fit the mould, you were excluded from this learning community.

The library's main attraction was the coveted balcony, which was reserved for future university students. This attitude permeated the rest of the library as students who were not scholars and who were not entering college or the workplace felt denied. On sunny days, these students could be found sprawled on the front lawn of the school, working together, collaborating, and discussing. There was no common place for all students to go to work, especially for those students who were disengaged from school and who were at risk of possibly dropping out.

Despite the fact that I was a strong student who conformed to the school's philosophy of learning, when I became a teacher, I became very interested in working with students who were the complete opposite of me when I was in high school. These students were disinterested and unmotivated, and it became a challenge for me to discover what could keep them connected to school.

When I was offered the position of student success teacher at my high school, I was very excited about the opportunity to meet with at-risk students and discover what was hindering their success in school. I knew that some of these students came from difficult home situations, and they had troublesome peer issues on a daily basis, so I really wanted a quiet place where these students felt comfortable confiding in me about their problems. I also realized that the majority of the students I would be talking to would be failing multiple subjects and might be embarrassed about their academic situation. They needed a safe place to strategize without having a lot of other students within earshot.

I went to the school's main office to see if there was any available space, but it seemed every office was being used. I was given a temporary room, but it was the one that was used for community

cynthia sargeant and roger nevin

SCHOOL LIBRARY 2.0: FROM THE FIELD

> "When a student....is struggling because of emotional, social, or behavioral issues, they do not go to the office – they go to the library learning commons."

> "With my new office, however, students are called down to the library, and they often arrive relaxed and in a more positive frame of mind because there are not negative connotations associated with the library."

counselors, and when they arrived at the school, their need for an office took precedence; I would be left homeless once again. I needed an office that was consistent, where students could find me if they needed me, that was somewhat private, and a place that did not have any stigma attached to it.

Roger Nevin, the teacher-librarian, mentioned he had an old book storage room in the library that was crammed with decades of discarded books and resources but if cleaned out could serve as a possible temporary office until I found something better. The idea of having a room that was all mine, that I did not have to share, appealed to me, and we soon got to work (with the help of some of my students) to make the book storage room a comfortable, safe location for students to meet with me. I was thrilled with the results. The room was not big, there were no windows, and there was absolutely no decor, but once I moved in a desk, a mini-fridge, a few posters, and a plant, it felt homey and comfortable.

The original idea of having an office in the library came by accident, but within a few weeks, I began to see the benefits of my location. Originally, when my office was next to the vice principal's office and I called a student out of class to see me, they arrived at my door anxious and defensive. Remember: Many of the students I served are deemed at risk because of behavioral issues—unable to cooperate within the confines of a classroom, they refused to do the work and so are unable to succeed. Therefore, from their past experiences, they immediately assume they will be getting a reprimand, a detention, or possibly a suspension.

Being "called down to the office" has a stigma attached to it, and it is embarrassing to them because everyone knows where they are going and speculate why. With my new office, however, students are called down to the library, and they often arrive relaxed and in a more positive frame of mind because there are no negative connotations associated with the library.

SUCCESS AND THE SCHOOL LIBRARY

Having my student success office in the library also had other advantages. If a student arrived and I was busy with someone else, Roger would invite the student to sit in the library's reading area and offer materials they might be interested in. These students do not often visit the library by choice, so having an opportunity to read quietly in the library while waiting for me was one way to encourage our at-risk students to begin to feel comfortable in this once foreign place. Not only did they become familiar with the library, but they discovered reading materials they actually enjoyed reading. We started to notice that these students began visiting the library, not to see me, but to sign out some of the materials they had been introduced to.

When I realized that combining the student success office and the library was attracting more students, Roger and I tried to devise other ways to keep the trend going. In spring 2007, I ran a leadership class twice per week so at-risk students could meet with a teacher and do team-building and collaborative-learning activities. We needed a space where students could move around for the dramatic activities but then have tables and chairs as well. In the library, there is a large seminar room that was a perfect spot for this class. Once again, the students in my school who were the most disengaged from school life were coming to the library for a positive program they really enjoyed.

When my at-risk students come to the library, either to meet with me or to participate in their leadership program, they see reading materials they are interested in. Whenever Roger goes book shopping for the library, I often tag along, making sure that he purchases books that will appeal to students of all reading levels. There are racks of Orca (high interest, lower reading level) right in front of the entrance, and current magazines are right next to them. Behind the front counter is the ever-popular Manga that is carefully selected upon the requests and recommendations of these students. Roger and I even host Manga parties at the library during lunch and invite a dozen disengaged and disinterested students who are enthused about Manga for pizza. This gives us an opportunity to get to know these students better. These are usually the first to check out recently purchased titles, and it shows them there are caring adults in the school who know what they are interested in and want to help meet their reading and social needs.

SCHOOL LIBRARY 2.0: FROM THE FIELD

Seeing these students excited and positive about the library clearly demonstrates that the libraries of today do not simply cater to the needs of the university-bound student. Rather, they can attract students of all abilities and attitudes.

Another way the library attracts disengaged students is with the computer lab at the center of the library. Students are welcome to use the Internet and check their e-mail outside of class time. The way young people use technology in their personal lives has changed dramatically over the last few years. From cell phones to iPods to Web 2.0 and such social networking tools as Facebook, students today use new technologies for communication and research. Schools are challenged to keep up with these tools. The library provides the best central location for creating the school's hub to support both students and teachers in the educational use of new technologies.

COLLABORATION/ CONNECTION

By integrating such technologies as Podcasting, simple video programs such as PhotoStory, and wikis into the curriculum, students become more engaged and pick up valuable 21st-century skills. Technology properly integrated into the curriculum especially engages at-risk students, who tend to be more active and are not passive learners. Also, assignments in these areas are more relevant to their personal lives—they are the MSN, iPod, and cell phone generation.

The collaboration between the student success teacher and the teacher-librarian is very important when successfully implementing new technologies into the curriculum. We provide guidance to classroom teachers by helping them enhance their curriculum. The assignments we recommend or design have been posted at connectingeducation.com, which we created together for other educators to use. I use my experience with at-risk students and Roger uses his technical skills to suggest the appropriate levels of technology integration. For example, instead of students just reading a novel, they can create a dramatic Podcast of the reading with music and sound effects. This Podcast can then be played in front of the class.

Together, we offer presentations to students on how to use these technologies, which means the classroom teacher does not have to learn how to use the new technologies. Our experience shows that students, especially at-risk ones, are more motivated when new technologies are used to enhance the curriculum.

MAKING THE LIBRARY STUDENT FRIENDLY

Resources and computers are not the only ways to make the library a student-friendly place. In order to attract disengaged students to the library, there must be a positive and encouraging atmosphere. School policy states that if a student returns a book after the due date, a late fee is imposed, but often Roger has a "sale," as he calls it, or he waives the fine altogether. Of course there are expectations for library behavior, but the rules are reasonable and not stifling. Roger occasionally needs to speak to students about their behavior, but he does so with a sense of humor and ensures that the encounter is not confrontational.

At Adam Scott, there is no balcony for the university-bound graduating students. Instead, all areas of the library are accessible to everyone, and there is an atmosphere of inclusiveness. All students are encouraged to succeed in a place that meets their needs. I do not know of any other school that has combined the student success office and the library, but here at Adam Scott, CVI it is a combination that really works.

Cynthia Sargeant (Student Success Teacher) and Roger Nevin (Teacher-Librarian) collaborate through the school library to support students, especially at-risk students. They are founders of boysread.com, which is a non-profit web site supporting male literacy and connectingeducation.com, which is a non-profit web site that provides resources for educators who want to connect education with young technology users. Cynthia Sargeant may be reached at *cindysargeant@gmail.com*; Roger Nevin may be reached at *connectingeducation@gmail.com*.

> "In order to attract disengaged students to the library, there must be a positive and encouraging environment."

FEATURE ARTICLE

SCHOOL LIBRARY 2.0: FROM THE FIELD

rethinking collaboration: transforming Web 2.0 thinking into real-time behavior

EDITOR'S NOTE: ONE OF THE CHARACTERISTICS THE TEACHER-LIBRARIAN WHO CONSTRUCTS A LEARNING COMMONS RECOGNIZES IS THE NEED TO BUILD CONSTANT CHANGE AND RETHINKING INTO THE PROGRAM AND OPERATION. STATIC PROGRAMS AND PATTERNS OF INTERACTION WITH OUR CLIENTS NO LONGER WORK BUT INSTEAD MUST BE FLEXIBLE AND EVOLVING CONSTANTLY IF THEY ARE TO BE RELEVANT. NOTE HOW SHEILA COOPER-SIMON REACHES OUT TO NEW IDEAS IN A CONSTANT REEVALUATION OF WHAT SHE IS CURRENTLY DOING AND HOW SHE CAN IMPROVE HER SERVICES.

sheila cooper-simon

By the time this article reaches the hands of readers, I know I will have changed my mind about some of what I say here. Sigh! Nevertheless, I like being a 1.0 thinker/writer sometimes, even though I know that my Web 2.0 side will heckle me, by virtue of those yet-to-emerge experiences that will change my ideas. This is a quandary I learned to articulate this year when I sat down with three Grade 5 students to explore how we would design an educational portal on a particular learning theme.

The online learning space we required at that time needed to reflect a number of aspects of learning. Much of it would need to be collaboratively constructed with Web 2.0 tools, but it also would need to include elements of Web 1.0 publishing. In other words, we needed at least two kinds of learning spaces: one that would allow students to be individuals who publish their work from their personal lens and another that would allow many students to construct knowledge through social networking processes. In kid language, they explained to me, this meant there were some things they would want to publish in a readable public space but that they did not want other students to have the power to change—"mess around with," or "wiki." They were thinking in particular of the sixth-grade—older, more powerful, and perhaps ominously more superior students. They did not even mention Grade 7s and 8s as a possible audience; it must have seemed so impossibly difficult to imagine the interface. So, in age-appropriate language, in response to my explanations to them of the components of Web 1.0 and Web 2.0 available to us, they pointed out some obvious ways to solve these problems and to achieve a learning commons that satisfied everyone's needs. I was suitably impressed by their problem-solving abilities because they hadn't actually ever heard before of a wiki, for example.

Fast forward a few months to my attendance of a presentation by cultural anthropologist Professor Michael Wesch (2007), at which he shared thoughts about his ongoing work with university students within an educational portal at Kansas State University. Perhaps the most remarkable piece of this experience was that listening to him from my perspective as a teacher-librarian working in two Canadian K–8 schools, very little of what he was saying about the potential of Web 2.0 to transform the way we teach and learn with students was particularly unfamiliar to me. How neat, I thought, that many of his comments and experiences echo mine so closely. In fact, his students'

SCHOOL LIBRARY 2.0: FROM THE FIELD

portal, albeit significantly more sophisticated, had many of the same components that the Grade 5s had thought they would need. Many other grade-school educators, as well as I, are exploring how to use educational portals, including how and when blogs, wikis, and discussion areas might be most effective with students at a number of grade levels. It seems to show a great deal of promise. Like Wesch (2007), too, experiences with students have led me to question the idea posited only a few years ago that there are separate generations of individuals who are either hardwired to be "digital natives" (younger people mostly) and "digital immigrants" (older folks certainly). I know that students are only a bit surprised these days when adults in the education system can talk with them comfortably about how they use their social networks like FaceBook and so on. Wesch's (2007) point is that new technologies emerge so quickly now that the playing fields have leveled, and increasingly the divide is closing. As I listened more, I thought, this correlation of ideas between his experience and mine is very tidy.

WEB 2.0 THINKING IS *NOT TIDY*

The tidy thinking didn't last very long. One single piece of Wesch's (2007) presentation startled me. He explained with articulate simplicity the problem of how large university classes and the current methodologies of delivering knowledge, frequently in a didactic style, can make students feel increasingly disconnected from the learning process and the instructor. This disconnect is complex. To understand it better, a quick visit to YouTube, searching *Michael Wesch*, will yield some explanations through a number of powerful videos he created with his students.

The parallel to this idea, with which I identified in thinking of my own profession and which startled me, was not that we deliver instruction in a didactic manner in school library programming but rather our problem is having *too many students* in proportion to one's role as a teacher within the teacher-librarian position. I am certain it has been said many times, university classes are too large and similarly teacher-librarian positions are fragmented and disproportionate to the numbers of students involved, not to mention significantly underfunded. But the simplicity with which he stated the problem and then went on to suggest that the thinking of Web 2.0 might in part solve both this instructional/learning problem and organizational/structural problem was astoundingly unique.

Apart from discussions about networked/hierarchical organizational cultures and authoritative/nonauthoritative information management, the essence of the problem in institutions is the way we *do* things on a daily basis, without deleting or changing existing, traditionally organized activity. The analogy in school libraries might be that how we manage our instructional time as teacher-librarians is sometimes dangerously akin to the problem of not weeding the print collection effectively.

Well, even though I use it and think I know what it is, how do we define Web 2.0 thinking so that it makes sense in relation to how we go about our collaborative work every day as teacher-librarians, I wondered? Here is a definition worth examining from Richard McManus (2005):

> Web 2.0 does appear to mean different things to different people, so you would be forgiven for still feeling confused about the term. Here are some of the definitions of Web 2.0 floating about: Web 2.0 = the web as platform; Web 2.0 = the underlying philosophy of relinquishing control; Web 2.0 = glocalization ("making global information available to local social contexts and giving people the flexibility to find, organize, share and create information in a locally meaningful fashion that is globally accessible"); Web 2.0 = an attitude not a technology; Web 2.0 = when data, interface and metadata no longer need to go hand in hand; Web 2.0 = action-at-a-distance interactions and ad hoc integration; Web 2.0 = power and control via APIs; Web 2.0 = giving up control and setting data free. While at first glance some of those definitions may be contradictory, we can distill from them certain characteristics of Web 2.0. Web 2.0 is social, it is open (or at least it *should* be), it is letting go of control over your data, it is mixing the global with the local. Web 2.0 is

> *I know that students are only a bit surprised these days when adults in the education system can talk with them comfortably about how they use their social networks like FaceBook and so on. Wesch's (2007) point is that new technologies emerge so quickly now that the playing fields have leveled, and increasingly the divide is closing.*

> "The analogy in school libraries might be that how we manage our instructional time as teacher-librarians is sometimes dangerously akin to the problem of not weeding the print collection effectively."

81e

about new interfaces—new ways of searching and accessing Web content. And last but not least, Web 2.0 is a platform—and not just for developers to create web applications like Gmail and Flickr. The Web is a platform to build on for educators, media, politicians, communities, and for virtually everyone in fact! Web 2.0 is all of the above things—don't let anyone tell you it's one or the other definition.

Do these definitions lead us all over the map, or is there a direction in Web 2.0 that we need to attend to? The new learning commons seems to be in fact the globalization of how we learn and are growing culturally. Libraries typically through time have engendered respect because they are trusted with the "keeping" of the knowledge. If that shift in the "keeping" has become more fluid, then it is likely that "reflecting" might possibly be closer to what the core piece of accountability is. Rather than backing away from that tremendously challenging piece, the simplicity of the lesson from Wesch (2007; one that should not be new to us) is that collaboration is the core of this process. It is within this complex role that school libraries need to reinvent their part. This is a tall order in light of the complexities that exist within school library structures, not the least of which is the marvelous uniqueness of each school library, a reflection of the uniqueness of each school culture. It is also complex in that one of the barriers to growth in libraries are stereotypes that follow the expected "finding" behavior that is apparently associated with libraries—in a time when, as Wesch (2007) and others say, we belong to a shift toward a "creating" culture, mirrored in Web 2.0. This, of course, translated to the role of library staff in Web 2.0 thinking, means not necessarily having a specific space in which to continuously work. Like everyone, library people do need a good landing space, and what a great central space they have from which to work in traditional library spaces. However, in addressing the needs of a school, it is no longer possible from one central space to address learning outcomes in a way that helps students to be prepared for a Web 2.0 world in which they must learn to create as well as find. Furthermore, that central space will actually be enhanced if the professional teacher-librarian leaves it sometimes to be in other physical and online learning spaces, much like the networking of Web 2.0. The library should continue to still be a physically viable place that represents a learning commons among many in and beyond the building. This is the shift from Web 1.0 thinking to Web 2.0 thinking. They both have a place in how we learn and collaborate.

WEB 2.0 THINKING AS AN APPROACH TO TIME MANAGEMENT

If you ask teacher-librarians to identify the most frustrating part of their job—the piece they wish they could change, the piece they feel least empowered to change—you will likely elicit a response that includes the word *time*. Frequently, too, an irony of the profession might be the inclusion of a strong protest about information overload. Who wants to admit that even as one of the "experts," you are finding it a little excessive and, quite frankly, a nuisance? Largely though, teacher-librarians will likely refer to the proportion of time in relation to the number of staff and students with whom they work and the gap that exists between the learning goals they would like to achieve and the reality of the number of human bodies with whom they work. Senge and colleagues' (2000) concept of the "creative tension" that exists between vision and reality comes to mind. There are two ways to cope: either reality can be pulled toward the vision or the vision can be adjusted to more closely match reality. Personally, I like visions, and some of the wonderful and current ideas for implementation in the professional literature about the role of teacher-librarians warrant some changing of reality to meet those aspirations. In keeping with Wesch's (2007) line of thinking, I suspect there are some changes to reality that may be possible within the thinking of Web 2.0. Some of that change might need to involve looking at different ways to use our time and perhaps to begin thinking about deleting some much-loved but outdated activities some of the time.

Teacher-librarians will sometimes claim that libraries, more frequently than any other part of school learning support, are the first victims of the proverbial axe when school budgets are tight. Sometimes this is very true. However, shortage of staff in relation to the job that needs to be done is certainly not unique to the school library profession. Fragmentation caused by less than full-time positions is rampant in many districts. On any given day, I can easily hook up with a classroom teacher, resource teacher, administrator, and, yes, even students with whom to lament how unfair it is they have so much to do within the given time frame. Such thinking, however, reeks of less than admirable human behavior, and there is whining involved. These are roads down which we do not want to go, however there is much to be learned from the positive side of these discussions with other professionals in schools, particularly those who feel their time allotment in proportion to students is limited.

How are they adjusting or getting past their frustrations? It is notable that among those solutions is the "thinking globally and acting locally" strategies, which are often the key points expressed. I think this means having a clear focus in any given school year on the big picture of the declared learning outcomes of the school and then choosing to focus on staff requests for collaborative activity within those goals as the key to time management. The more closely the inquiry units and information literacy lessons match with learning goals of the school and curriculum outcomes, as agreed upon by the participating collaborators, the more time should be spent with those particular students and teachers. As expressed by Web 2.0 philosophy, it is the local interface of goals that should become the construction or work area, a kind of learning commons.

The most immediate argument that might be posed to this decentralized approach is whether or not one is fulfilling the profile of a school library program that meets the needs of all students and staff. Will there be evidence that continuous, systematic attention is being paid to all grade levels? The quick answer to that is to think of the school library program as a "spiraling curriculum." It is not possible in a time frame of one school year to fit in collaboratively planned, good quality work with teachers and students in classrooms, computer labs, or school libraries that reflect the deeper thinking processes that are recommended by Harada, Loertscher, Kuhlthau, and Stiggins, and countless other researchers. One needs to prioritize the tasks and allocate time in accordance. Certainly one needs to keep touching base with staff with whom less work is being done in a given year, perhaps spending time with them for short lessons, support, or simply discussing what they are doing. However, if they feel they are achieving the same goals independently from the school library program within a given year, then all that is necessary is conversation about how their learning activity connects with what you are doing collaboratively with other students and staff. The focus of energy needs to return always to attending to the learning activities that include the best intersection of all stakeholders' goals. These goals are quite different from one school culture to another, a fact I can attest to through my daily experiences within two schools.

As a teacher-librarian making decisions in this big picture to perhaps dramatically decrease or delete the amount of time spent, for example, on collecting resources or reading to children, it becomes our job to converse about why we aren't doing those things in a given year. Maybe we need to offer solutions to those questions, too. A solution to the "why aren't you reading to students?" problem might be to develop an online book club that is mostly run by students or have the library technician or classroom teacher read to younger children in the library setting. These solutions usually involve an investment of time in the front end, but much like in the Web 2.0 world, they take on a life of their own that gives ownership of the learning to all participants.

The work that is accomplished with students and staff in any given school year certainly does need to be systematically organized for accountability. This means that several layers need to be considered in any given school year. In my province, most of those layers are already designed for me through curriculum and curriculum support documents. It is my job to figure out how to put them on a grid that fits with what I believe I can achieve with classroom teachers and their students. For example, my province has a continuum of information and communication technology learning outcomes, frameworks for all of the curriculum areas, and support documents that address such considerations as inclusive education and multilevel classrooms. I can choose to adopt a specific inquiry model, such as Alberta's Focus on Inquiry model, to use with teachers who wish to use a model, or I can use a resource, such as Kulthau, Maniotes, Caspari's newest publication, *Guided Inquiry: Learning in the 21st Century* (2007). The resulting grid that guides the instruction and assessment of the teacher-librarian for the school year becomes a guide and communication tool. Like a Web 2.0 tool, it is open to being changed and reconstructed throughout the year in response to the learning needs that are identified by all the stakeholders. This is particularly true as technologies and ideas surrounding new technologies are so quickly changing the way we can implement instruction with information literacy or new literacies.

ASSESSING STUDENT LEARNING OUTCOMES, ASSESSING INFORMATION LITERACY: WHEN AND WHY?

One of the most challenging parts of making a shift to being part of the

> "The library should continue to still be a physically viable place that represents a learning common among many in and beyond the building."

> "I think this means having a clear focus in any given school year on the big picture of the declared learning outcomes of the school and then choosing to focus on staff requests for collaborative activity within those goals as the key to time management."

> "The focus of energy needs to return always to attending to the learning activities that include the best intersection of all stakeholders' goals."

organized disorder of Web 2.0 behavior is deciding how and when it is most valuable for the teacher-librarian to participate in assessment and evaluation of student learning. Within a unit of study that one has presumably planned through the backward design process in collaboration with a classroom teacher, it seems necessary to complete the process by assessing the student learning together at the completion of the learning cycle. This is time-consuming and is often the piece that makes it necessary to choose to work with one classroom over another. These are not easy decisions to make and, again, can probably only be partly addressed through the idea of spiraling one's activity. Staying visible to those students with whom one is not working on a larger unit can possibly be addressed through single visits to classes in the computer lab, for example, in which one spends time chatting and reviewing how their Web searching is progressing. Sometimes, it is even most effective to introduce a new online resource to students and a teacher through a quick drop-in like this in which one knows they will share that new resource with other students or teachers. The logic of choosing this kind of behavior is to move away from the idea of finding information in response to questions and instead provide ideas to learners for relevant use of information and presumably for them to share with others.

Because there should be a continuum of brief and more comprehensive assessments within any given day, it can be valuable to allocate a couple of days per week at least to formally plot onto a spreadsheet what has been assessed. It can be effective if choices of which assessments to plot are made randomly or targeted, as long as one is aware of the purpose of the data collection and analysis. Depending on what kind of analysis works best for the individual teacher-librarian, this usually helps to provide a sense of the patterns of learning that are occurring. This becomes the fodder for conversation and significantly lessens the sense of isolation that sometimes occurs within fragmented positions. Perhaps one of the pieces of Web 2.0 philosophy that most aptly applies to assessment is that we are interdependent in our observations of student learning. Teacher-librarians have more opportunity to communicate with other teachers than almost anyone else on staff by virtue of their collaborative positions.

WHY YOU MAY OR MAY NOT NEED A SCHOOL LIBRARY WEB SITE

Is having a school library web site important in light of Web 2.0 thinking, where networked, overlapping spaces take precedence? Do we still need to provide an organized hierarchical Web 1.0 directory of good sites to visit for general reference on a school library web site? Even though Google and a number of other search engines do this extremely well, most teacher-librarians likely believe it is still necessary. However, the issue of maintaining it over time becomes a problem. If one is focused on student learning outcomes through collaborative instruction and assessment, when is there time to maintain a good web site? Again, the alternate way of managing this might be to change the ownership of the site. By creating a learning commons for the school library, perhaps inviting a group of students to assist with the design, one can invest time in the beginning of a new school library web site and then, over time, move to the role of managing editor. Most students are responsible about contributing to a common learning space, especially if they know an adult is checking it regularly.

THE SHIFT

The shift of thinking that is exemplified by the concepts of Web 1.0 and Web 2.0 is endlessly fascinating as it captures and reflects the essence of social change. It is impossible to predict what will work well for students in these learning forums. However, school libraries have a wonderful bird's-eye view of how the pieces mesh because we see so many students. Perhaps the problem, *too many students*, holds the clues to the solution. Maybe the forest is there among the trees. At any rate, it is both a delight and a responsibility to listen carefully to how it unfolds.

REFERENCES

Kulthau, C., Maniotes, L., & Caspari, A. (2007). *Guided inquiry: Learning in the 21st century.* Westport, CT: Libraries Unlimited.

McManus, R. (2005). *What is Web 2.0.* Retrieved July 19, 2008, from http://blogs.zdnet.com/web2explorer/?p=5.

Senge, P., Cambron-McCabe, N., Lucas, T., Smith, B., Dutton, J., & Kleiner, A. (2000). *Schools that learn.* New York: Doubleday.

Wesch, M. (2007). *A vision of students today* [Electronic version]. Retrieved July 18, 2008, from www.youtube.com/watch?v=dGCJ46vyR9o.

Sheila Cooper-Simon is currently a teacher-librarian in Winnipeg, Manitoba, Canada. She has a M. Ed. in Teacher-Librarianship from the University of Alberta and is pursuing an M. Ed. in Educational Administration from University of Manitoba. She is a former President of the Manitoba School Library Association and is currently a Councillor of the Canadian Association for School Libraries. She can be reached at *simon@mts.net*.

> Because there should be a continuum of brief and more comprehensive assessments within any given day, it can be valuable to allocate a couple of days per week at least to formally plot onto a spreadsheet what has been assessed.

Part IV:

Technology and the Learning Commons

For some time, the role of technology in education has been debated at the same time it has been rapidly growing. Fear of student behavior on networks as well as the presence of the porno industry, hacker mentality, scams, and predator behavior has cause limitations to access and simultaneous calls for Internet safety initiatives.

In the proposals for a Learning Commons, Loertscher, Koechlin, and Zwaan call for the division of computing into two segments for each school and district – those of administrative computing and instructional computing. Administrative computing including such things as attendance, budgets, grades, and personal information should obviously be locked down and protected. However, instructional computing should be more open since it serves student information and opens the door to the wonderful world of Web 2.0 tools and cloud computing.

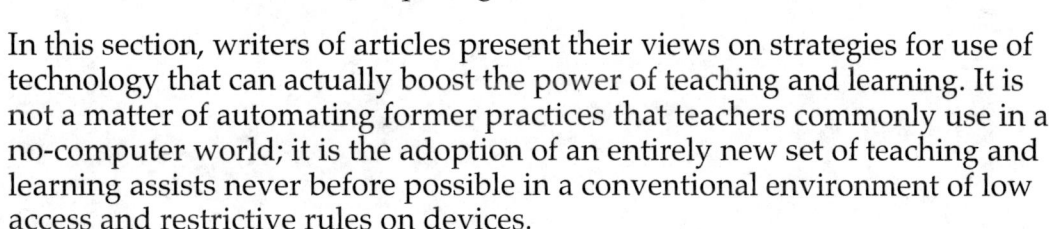

In this section, writers of articles present their views on strategies for use of technology that can actually boost the power of teaching and learning. It is not a matter of automating former practices that teachers commonly use in a no-computer world; it is the adoption of an entirely new set of teaching and learning assists never before possible in a conventional environment of low access and restrictive rules on devices.

The lead article by Marcoux and Loertscher is actually a compilation of a wide perspective generated by many contributors as the group reconsidered the various contributions of technology to teaching and learning. Other articles investigate new strategies in network infrastructure, the use of Google Apps and cloud computing, and the inclusion of social networking. All the articles stress the probabilities and possibilities of increased impact on student engagement in learning.

FEATURE ARTICLE

Achieving Teaching and Learning Excellence With Technology

"Now, educators face the second decade of the 21st Century with seemingly unlimited ways technology can influence what we do."

ELIZABETH "BETTY" MARCOUX AND DAVID V. LOERTSCHER

Since the very first introduction of a Commodore Pet, TRS 80, and the Apple II microcomputers beginning in 1977, billions of dollars have been spent chasing a dream about the effect of technology on teaching and learning.

Some thirty years later, the exponential development of powerful devices, networks, the Internet, software, and now Web 2.0 keeps the effect dream alive, but the results somewhat elusive.

By 1977, teacher-librarians and audiovisual specialists were busily integrating various multimedia into school libraries and audiovisual collections. Few people could imagine at the time or embrace the seemingly outlandish claims for devices doing work within a whopping 48K or the gigantic 64K machines.

Now, educators face the second decade of the 21st Century with seemingly unlimited ways technology can influence what we do. Simultaneously, children and teens of this generation have enthusiastically embraced technology for social networking and content creation purposes but have failed or not been allowed to extend their technology expertise over into their academic pursuits.

DOES TECHNOLOGY MAKE A DIFFERENCE?

There are a litany of reasons from research and the professional literature detailing reasons why technology does not fulfill the often bloated expectations, but we would like to focus on some observations of our own.

For decades, a popular approach to technology is what we would refer to as device driven applications. When a new technological device hits the market, we begin by trying to discover its characteristics and then imagine how those characteristics could be used in teaching and learning. The same goes for software and in particular Web 2.0 apps. Here is an application, here is what it does, and here are some ideas of how it could be used. All of us respond to lists of apps or smackdowns whereas we are entertained and dazzled by endless gizmos, a new tool or toy, and neat new discoveries.

Such sessions, while fun and exciting, often leave one with an overwhelming feeling of confusion and inadequacy. What should I buy? What 25 apps should I investigate next? How do I keep track of what does what?

THE LEARNING TO TECHNOLOGY APPROACH

The authors and contributors of this article set out in the opposite direction. We call it the learning to technology approach. Begin with a learning problem, diagnose that problem, and then prescribe one or several excellent tools that will work to solve that "problem." We could also say begin with best practices you want to achieve and then fit the tool to that challenge. Become the doctor, not the pharmacist.

In our lists we grouped the many characteristics of both devices and software into six major categories, followed by a focus on the organization to deliver those teaching and learning benefits. Our argument for this arrangement centers on excellence in teaching and learning—the idea that technology use must be tested for both individual and group growth in a global world. Thus, we end the listings under each characteristic with encouragement to collect data on that characteristic and report it widely. We give the major reason for this in our conclusion: Actual data and evidence collection is beyond the scope of this article. There are sources that will help in that focus, so first focus on results and then develop techniques to measure effect.

We are quite certain every reader will be able to add both items under each characteristic as well as find examples that have produced results with real learners in their schools. Perhaps this list can be the foundation of a major conversation in professional learning communities and in technology-focused professional development. Here is the list:

EFFICIENCY (STUDENTS AND ADULTS WORK TOGETHER SMARTER)

■ Both students and adults build organizational skills to handle the world of information and technology and turn it into a foundational tool to boost their learning. Examples:
• Shared calendars helped everyone meet deadlines.
• Group project roles, deadlines, and responsibilities led to more efficient collaboration.
• Individuals built their own information spaces that gave them power over the onslaught and juggernaut of the Internet.
• Communication across the room, the school, the community, and the world allowed for projects of a larger scope.
• Writing and research were enhanced by tracking bibliographies, quotes, sources, note taking, and editing.
• Customized search engines that probed in-depth into information and documents provided better and more relevant searches.
• Gadgets, widgets, RSS feeds, and alerts connected learners automatically to needed content, news, blogs, and people.
• Collaborative construction of documents, presentations, and creative works were done in real time.
• Collaborative creation of works that could be tracked, monitored, edited, developed, and assessed by students and teachers over time had a direct effect on quality of thinking and sharing.
• Collaborative linking and sharing of resources brought a wider variety of information into the pool for various learners.
• Opportunities to deliver products that represented learning also reflected the students' learning style.

■ Technology assists provides everyone the opportunity to create professional-looking presentations, products, reports, videos, audio, and mashups.
• Technology assists benefit all learners whether they are gifted, challenged, or have varying learning styles.
• True differentiation to boost productivity becomes a reality with a variety of tech tools.
• Work time with many applications streamlines searching, building, and constructing so more time can be spent in thinking, reasoning, and sound decision-making as well as analysis, synthesis, and reflection.
• Student-centric technology allows customization and provides the ability to tailor learning to individual student needs.

Evidence of the effect of technology on efficiency is collected and reported widely.

MOTIVATION TO LEARN

■ Variety and novelty of a technology or a new technique piques attention, motivation, and engagement. Examples:
• A Geek squad of students taught the entire school new technology in one week.
• A class was offered for 1/4 credit per year for students to tend the help desk, participate in training for computer repair, and provide a regularly published "Tips" newsletter for the school community. Previously "bored" students suddenly had a purpose in coming to school.
• A plethora of choices of devices, tools, and techniques facilitated wide choice in product creation.

■ Use of technology boosts engagement of a higher percentage of learners as compared to textbook lecture strategies. Examples:
• Social networking skills began to extend toward a new culture in academic skills.
• Using blabberize.com, elementary students were able to share information from their "animal reports" in a much more dynamic way.

■ Real problems and issues boost both engagement and deep understanding. Examples:
• Using a handy miniature video camera, students were excited to interview law enforcement personnel about a school disaster plan they were creating.
• Projects across borders brought a sense of community, sharing, and learning across cultures.
• Presentation tools encouraged students to reach out to adults in their communities and work to solve real-time problems.

■ Personalization of learning allows all learners to make choices, to take command of their own learning, and to capitalize on their personal strengths.
• Many tech tools allow experimentation and playing with ideas thus nurturing creativity and inviting innovation.

■ Presenting, performing, and sharing with peers stimulates the quality of products, work, and the desire to participate fully. Examples:
• Young children each created a "slideshow," which they presented to parents.

> "Begin with a learning problem, diagnose that problem, and then prescribe several excellent tools that will work to solve that 'problem.'"

> "Work time with many applications streamlines searching, building, and constructing so more time can be spent in thinking, reasoning, and sound decision-making as well as analysis, synthesis, and reflection."

> Reflection Question: What does "deep understanding" look like when technology tools are part of the equation?"

• Presentations by individuals and groups were available for simultaneous sharing, analysis, synthesis, and assessment.

• Student presentations became more sophisticated using mash ups and a variety of technologies.

■ Collaborative spaces extend beyond purely social concerns toward, constructing, sharing, and motivating others, and present opportunities to not only raise student interest in learning, but also allow them to grow from each other's insights. Examples:

• Many tools for forming reading networks, sharing spaces and encouraging, critical analysis promoted a high interest in reading, writing, and enjoyment of multiple genres in numerous small to large group environments. Such groups took on a life of their own.

• Collaborative spaces raised "students' level of concern" and encouraged them to collaborate when they could read thoughtful responses of their peers using their literature circle wikis.

Evidence of the effect of technology on motivation to learn is collected and reported widely.

DEEP UNDERSTANDING

• Certain tech tools allow for both individual and collective knowledge building in conjunction with the process skills that are being developed. Examples:

• A group of various non-native English speakers created varying pictorial representations to understand science and social studies concepts.

• Older students developed electronic resources for younger students and learned about child development at the same time.

■ When content, substance, original and creative thinking, logic, and reflection are the focus of assessment over the admiration of glitzy and slick presentations, the focus of presentations and projects turns toward academics. Example:

• Students began to notice that adults were more interested in the content of their presentations rather than splashy tech features. When one group got a higher grade for their presentation than did a special effects presentation with little substance, the word got around: Know your stuff!

■ Real and authentic problems or issues combined with inventive uses of technology boost sustained interest, deep understanding, and engagement. Examples:

• A collaborative study of school violence and bullying utilizing many Web 2.0 tools expanded student interest in taking major action both in their school and across schools in their community.

• Lead teachers designed a Web 2.0 environment to facilitate the study of a Professional Learning Community. They modeled the building of deep understanding of professional concerns as they learned how to use the technique with their students.

• Live television coverage and web-streaming of school events allowed students to learn the production process and provided an opportunity to share their voice with the community.

• Web-based student publications—a school newspaper (or literary journal)—provided dynamic environments for student journalists and writers to hone their skills in a 24/7 environment while also learning about the editorial process and the importance of authority.

• Technology can change the way we engage with content and/or actually add to content. Example:

• The use of GPS technologies had an effect on how students saw the geographic and political world. One class proposed a novel two-state solution between Israel and the Palestinians, and sent their proposal to the parties involved.

■ Deep understanding is stimulated by the delivery and interaction with resources via technology that can be done no other way. Example:

• Video of how a human heart actually works enhanced deep understanding along side two-dimensional diagrams, descriptions from the printed page, models of the heart, and data from various sensing technologies.

Evidence of the effect of technology on

deep understanding is collected and reported widely.

LEARNING HOW TO LEARN (21ST CENTURY SKILLS)

- Certain tech tools allow for building both individual and collective learning skills competence in conjunction with the content knowledge and deep understanding that are being developed. Example:
  • Students used mind mapping tools such as Mindmeister, Gliffy, bubble.us, and MyWebinspiration to build ideas about people, places, and issues as a small group; then they jigsawed to understand varying interpretations.

- Multiple literacies that involve new techniques and new methods of analysis evolve as new competencies are required for new tools and applications. Example:
  • Adults noted that social networking required new participatory skill sets as Jenkins describes in *Confronting the Challenge of Participatory Culture: Media Education for the 21st Century* (2006). The adults focused on those skills across a semester measuring progress across time.

- Learning how to collaborate as a creative and skilled group member can be done across blogs, wikis, and back channels. Examples:
  • Before doing their collaborative research, students, classroom teachers, and teacher-librarians used a Google Spreadsheet to create suggested rules about team responsibility during research. After brainstorming, the entire group looked at the suggestions on the spreadsheet as a whole and then built a common set for the project at hand. After the project, the class reflected on progress made in collaboration.
  • When students worked collaboratively online and teachers were able to view the history of their documents, all participants of the collaboration became accountable for their contribution and students took ownership of their collaborative products.

- Finding quality information on the Web and sorting through the voices of who is saying what to me for what reasons and for what gain is an essential life skill. Examples:
  • Collaborative teams searched for and evaluated the credibility of web sites in order to defend their use in a joint project.
  • Instruction about how to evaluate a student's product regarding the quality of information included was the theme of a professional development opportunity session with follow-up three months later.

- Learning the research process and other information literacy skills is a part of building content knowledge. Examples:
  • Students used tech assists to help in the research process including search engines, note taking, attribution, analysis, synthesis, writing, presenting, and reflection.
  • Students used various tech tools to do collaborative research in addition to building individual knowledge. They could discuss what they knew as individuals as well as an entire group.
  • Students with a variety of abilities were able to research using differentiated web resources and combined learning in a group project.

- Content and media creation is available to everyone in the school community and the principles of intellectual property can be explored. Examples:
  • The school created its own internal "YouTube" to showcase the best of the best creative and academic products.
  • The entire school community learned how to strike a balance between content creation and the ethics associated with intellectual and creative property.
  • The entire school community used the Creative Commons and other content repositories to build and share content that can be repurposed and shared.
  • Expanded ideas of fair use were promoted such as those included in "The Code of Best Practices in Fair Use for Media Literacy Education," available at http://www.centerforsocialmedia.org/resources/publications/code_for_media_literacy_education, and Joyce Va-

> "Finding quality information on the Web and sorting through the voices of who is saying what to me for what reasons and for what gain is an essential life skill."

> "Tools and options that appeal to various learning styles are offered as a matter of accomplishing any task or assignment."

lenza's article "New Fair Use Code of Practice: A Call to Action" available at http://www.schoollibraryjournal.com/blog/1340000334/post/1200036320.html?q=fair+use.

■ The research skills using information literacy models are expanded into the larger realm of 21st century skills. Examples:
• During a research project, media literacy, creative thinking, and critical thinking were stressed.
• Research skills regarding guided inquiry were matched to new standards at both state and national teacher-librarian conferences.

■ Tech tools can be selected with specific 21st century skills in mind (critical thinking, problem solving, collaboration, information literacy, ICT literacy, flexibility, innovation, creativity, global competence, and environmental literacy). See the MILE Guide from the Partnership for 21st Century Learning (2009). Example:
• In a study of African countries, the class used Google Maps to peer into the actual geography, culture, with a sense of real time exploration to enhance true global understanding that could not be done with print or other multimedia.

Evidence of the effect of technology on learning how to learn is collected and reported widely.

CREATIVITY AND CONTENT CREATION

■ The web opens the flood gates to individual and group creation of serious content. Examples:
• Students wrote for Wikipedia and published serious reports/projects on YouTube and other sites hoping they would go viral.
• Students used various technologies to present information about how to not only find information, but how to use information effectively on Facebook.
• In a global learning experience, groups of students were formed including members from other countries. In spite of language difficulties, joint mashups were made.

• Students took folktales from various cultures in their global groups and re-wrote them with a different cultural perspective. Video re-enactments of both versions were made and shared.

■ Both formal learning and informal learning combine to build the creative and the serious self. Example:
• A band instructor encouraged self-creation and performance both at school and at home via music generation software and performance that went global.

Evidence of the effect of technology on creativity and content creation is collected and reported widely.

INCLUSION OF DIFFERENT TYPES OF LEARNERS

• Various assistive devices provide opportunities to those with physical disabilities such as low vision, deafness, and limited mobility. Examples:
• The Geek squad produced a Jing video about text-to-speech software on various devices and played the video to the entire school with a challenge to teach the skill to family members or neighbors who might benefit.
• Text-to-speech software allowed those with low vision or are completely blind to listen to texts.
• Skype and other communication technologies allowed children and teens to communicate in real time across the world.
• IVC (Interactive Video conferencing) enabled deaf learners to communicate by signing.
• Chronically ill students continued to be involved in school projects by using online technologies.
• Both adults and students organized class tests, assignment calendars, and student folders using Web 2.0 tools so disasters such as hurricanes, epidemics, and power outages did not stop school.
• Various tools such as Kindle were analyzed and tested by students for their usefulness as inclusion.

■ Teacher-librarians build expertise along with technology staff of the school to be masters at wise and effective use of technology in order to boost the quality of teaching and learning. Examples:
• Teachers with an instructional problem in their classroom often came to the teacher-librarian for recommendations on just the right Web 2.0 tool to use together in solving the problem.
• Technology programs at conferences were regularly attended as professionals incorporated the best ideas encountered into their schools for students with learning, physical, and attitudinal challenges.

■ Second language learners benefit with tools that build reading, writing, and sharing skills, as well as tools that bridge the language gap such as visuals, mind mapping, vocabulary boosts, and translation tools. Examples:
• A class used BookFlix to read children's books in English and Spanish.
• Databases and other online resources were available in a variety of languages for the students' use during a project they were working on.
• Individualized tutoring technologies such as "My Reading Coach" helped children learn to speak and read English.
• ESL students found technology tools to be more forgiving of their mistakes and thus were more motivated to return to them for further language instruction.

■ Tools and options that appeal to various learning styles are offered as a matter of accomplishing any task or assignment. Examples:
• The ability to collaborate with others helped students learn from one another in a less threatening environment.
• Students were able to continue work at home or in other locations via the Internet, using online document producers, flash drives, and e-mail.
• Class presentations were created in a variety of formats, combining student skills and experiences.

■ Non-traditional learners are provided the tools needed to both engage them and include them in the learning at hand. Examples:
• Shy but articulate students who did not speak out in class suddenly bloomed in

online collaboration and discussions.
- Conversely, students who were excellent speakers but poor writers applied their talents using a variety of media as they completed assignments, all accepted by the teacher.
- Evidence of the effect of technology on inclusion of different types of learners is collected and reported widely.

ORGANIZATIONAL CONCERNS AND SUPPORT

■ The technology leadership team of the school and district include tech directors, administrators, teacher technologists, teacher-librarians, representatives of the faculty, and representatives of the students and their parents. Examples:
- When doing renovation, teacher-librarians and the technology staff were housed in the same facility and therefore could work even closer together.
- Tech savvy teachers along with the tech director and the teacher-librarian formed a professional learning community charged with maximizing the effect of technology on teaching and learning.
- The school and district provide equitable access to networks and devices as well as access to information and resources that promote excellence in teaching and learning.

■ Robust wireless access is available throughout the school, in particular the learning commons. Example:
- The IEEE standard 802.11g was replaced by 80211n that provided ubiquitous access to an entire school community.
- Emergence from a school library and computer lab into a Learning Commons concept is an important aspect of moving to the center of teaching and learning.

■ Each student is equipped with a device of choice to access materials and resources 24/7/365. Example:
- In anticipation of the installation of an 802.11n standard network, the district committed to open each school's network to any and all devices that various students owned personally in addition to those purchased by the school.

■ Expertise to assist with networking, trouble shooting, peer tutoring, and developing technological expertise includes both adult experts and peers in a "you help me, I help you, and we all learn together" atmosphere. Examples:
- The school geek squad of students had a mission to be of assistance to every teacher and student in the building. They prided themselves in being able to teach the entire school a new application in a matter of hours or days.
- Technology leaders developed a wide range of "experts" within the system and readily facilitated the expansion of teaching and learning across the learning community.
- Entire departments of tech "experts" were organized at the district level to assist with technology concerns encountered in teaching and learning.

■ Professional development in technology focuses on long-term sustainability of both best practices using technology and experimental applications and strategies that affect teaching and learning.
- Students are considered partners in the development of technology systems, practices, policies, as well as dissemination, and other issues related to the spectrum of utilizing technology in education.
- The Learning Commons staff has the responsibility to provide the rich and high quality information environment in which learners can thrive. Such an environment requires a substantial financial commitment to provide the information and media plus the networks and technologies to access it.
- The foundational principles of intellectual freedom extend to networks, devices, tools, and information.
- Partnerships with other organizations, the community, consortia, and granting agencies provide the wherewithal to implement the constant change and improvement required to keep pace with technology.

■ Teacher-librarians along with other technology professionals collect data about the effect of technology on teaching and learning over and above reports concerning networks, computers, and spending on software/databases. Example:

"Professional development in technology focuses on long-term sustainability of both best practices using technology and experimental applications and strategies that affect teaching and learning."

"There is a great opportunity to have major influence on the drive toward excellence and demonstrate there is power in the results achieved by both individuals and groups; both adults and learners."

• The teacher-librarian recommended several Web 2.0 solutions to solve the problem of students not writing enough or of high enough quality about what they were reading in literature or on other academic topics. The data on the use and effect of those tools was collected and disseminated widely.

CONCLUSION

During the era of No Child Left Behind (2002) excellence was measured in terms of performance on one or several standardized tests. In the new era of Race to the Top (2009) with money flowing presumably toward innovation and multiple measures of achievement, there is a new opportunity for teacher-librarians and teacher technologists if they realize they have the power through technology to move into the center of teaching and learning. There is a great opportunity to have major influence on the drive toward excellence and demonstrate there is power in the results achieved by both individuals and groups; both adults and learners.

Instead of a group of technologies and apps waiting to be used, consider the number of learning and learner challenges for which particular applications are especially good in making a difference.

We suggest that you focus first on the learning problem or challenge; on the problem at hand; that challenge faced. Then and only then introduce particular technologies that you have some confidence will work and that have succeeded in the past. However, do not be afraid to take a risk with newer and more exciting technologies and applications that come down the pike. It all keeps getting better and better; the opportunities get greater; and the track record easier and easier to recognize, document, and report widely.

NATIONAL STANDARDS DOCUMENTS

From AASL

Standards for the 21st-Century Learner: http://www.ala.org/ala/aasl/aaslproftools/learningstandards/standards.cfm

From ISTE:

NETS (National Educational Technology Standards) for Students: http://www.iste.org/Content/NavigationMenu/NETS/ForStudents/2007Standards/NETS_for_Students_2007_Standards.pdf

NETS for Teachers: http://www.iste.org/Content/NavigationMenu/NETS/ForTeachers/2008Standards/NETS_T_Standards_Final.pdf

NETS for Administrators: http://www.iste.org/Content/NavigationMenu/NETS/ForAdministrators/2009Standards/NETS-A_2009.pdf

From the Partnership for 21st Century Learning:

Partnership for 21st Century Skills. Route 21: http://www.21stcenturyskills.org/route21/

MILE Guide (Milestones for Improving Learning & Education) http://www.21stcenturyskills.org/documents/MILE_Guide_091101.pdf

MILE Guide Chart: http://www.21stcenturyskills.org/images/stories/otherdocs/p21up_MILE_Guide_Chart.pdf

MILE Guide Online Assessment: http://www.21stcenturyskills.org/index.php?option=com_wrapper&Itemid=95

MILE Guide Workshop Kit: http://www.21stcenturyskills.org/images/stories/otherdocs/p21up_MILE_Guide_Workshop_Kit.pdf

The Intellectual and Policy Foundations of the 21st Century Skills Framework: http://www.21stcenturyskills.org/route21/images/

International perspectives:

Co INCITS (International Committee for Information Technology Standards): http://www.incits.org/new_stds.htm

From the U.S. Government:

National Educational Technology Plan: https://edtechfuture.org/

Common Core Standards Initiative: http://www.corestandards.org/

STATE STANDARDS DOCUMENTS (SAMPLES)

• Arizona Technology Standards: http://www.ade.state.az.us/standards/technology/
• Massachusetts Technology Standards: http://www.doe.mass.edu/edtech/standards.html
• Washington Educational Technology Standards: http://www.k12.wa.us/edtech/techstandards.aspx

Check out your own state or Provincial government for technology standards.

OTHER INFORMATION

14 Ways K–12 Librarians Can Teach Social Media: http://www.techlearning.com/editorblogs/23558

21st Century Skills Support Systems Implementation Guides (draft): http://www.weareteachers.com/web/407596/discussit

30+ Alternatives to YouTube: http://www.freetech4teachers.com/2009/06/30-alternatives-to-youtube.html

Boss, S., & Krauss, J. (2008). *Reinventing Project-Based Learning.* Washington, DC: ISTE.

Center for Safe and Responsible Internet Use: http://csriu.org/

Challenge Based Learning: http://ali.apple.com/cbl/index.html

Consortium for School Networking. (2009, May). Leadership for Web 2.0 in Education: Promise and Reality: http://www.cosn.org/Default.aspx?id=85&tabid=4189

Digital Transformation: A Framework for ICT Literacy—A Report of the International ICT Literacy Panel. Princeton, NJ: Educational Testing Service. http://www.ets.org/Media/Tests/Information_and_Communication_Technology_Literacy/ictreport.pdf

Digiteen: A web page for a digital citizenship group project between Qatar Academy, Westwood Schools in Camilla, Georgia USA, and Vienna International School in Vienna, Austria. Available at http://digiteen.wikispaces.com/

Fitzgerald, M. A., Orey, M., & Branch, R. M. (annual). *Educational Media and Technology Yearbook*. Santa Barbara, CA: Libraries Unlimited.

Free the standards: http://www.schoollibraryjournal.com/blog/1340000334/post/1590046559.html

From Now On: The Educational Technology Journal by Jamie McKenzie: http://fno.org/

Google Co-founder Sergy Brin Wants More Computers in Schools: http://latimesblogs.latimes.com/technology/2009/10/sergey-brin-put-computers-in-schools-.html

Hendron, J. G. (2008). RSS for Educators: Blogs, Newsfeeds, Podcasts, and Wikis in the Classroom. Washington, DC: ISTE.

Horizon Project: http://www.nmc.org/horizon

Horizon Report Wiki: http://horizon.wiki.nmc.org/

Jenkins, H., et al. (2006). "Confronting the Challenge of Participatory Culture: Media Education for the 21st Century." MacArthur Foundation: http://www.projectnml.org/files/working/NMLWhitePaper.pdf

Johnson, L., Levine, A., & Smith, R. (2009). The 2009 Horizon Report (Trends report on technology): http://www.nmc.org/pdf/2009-Horizon-Report.pdf

Jonassen, D. H., et al. (2007). *Meaningful Learning with Technology*. 3rd Edition. Upper Saddle River, NJ: Prentice-Hall.

Kindle Lightens Textbook Load, but Flaws Remain: http://www.google.com/hostednews/ap/article/ALeqM5jEb4TakU-nHP-ECdCZDjv7C5ejUkAD9BAE7QG0

Leading the Way to Transforming Learning with 21st Century Technology Tools: http://www.tomorrow.org/speakup/pdfs/SU08_findings_final_mar24.pdf

Learning Tools: http://c4lpt.co.uk/learningtools.html

Li, C., & Bernoff, J. (2008). *Groundswell: Winning in a World Transformed by Social Technologies*. Boston: Harvard Business Press.

Macarthur Foundation. *The Future of Learning*: http://mitpress.mit.edu/books/chapters/Future_of_Learning.pdf

Mindset: http://mindsetonline.com/whatisit/about/index.html

National Education Technology Plan Development Work: www.edtechfuture.org.

November Learning by Alan November: http://novemberlearning.com/index.php?option=com_frontpage&Itemid=1

November, A. (2008). *Web Literacy for Educators*. Thousand Oaks, CA: Corwin Press.

Pitler, H., et al. (2007). *Using Technology with Classroom Instruction that Works*. Alexandria, VA: ASCD.

President Obama, Secretary Duncan Announce Race to the Top: http://www.edgovblogs.org/duncan/2009/07/president-obama-secretary-duncan-announce-race-to-the-top/ and http://www.thenewamerican.com/index.php/culture/education/1509

Research Shows Schools Making Small Progress Toward Technology-Rich Environments: http://thejournal.com/articles/2009/07/02/research-shows-schools-making-small-progress-toward-technology-rich-environments.aspx

Rose, D., & Meyer, A. (2002). *Teaching Every Student in the Digital Age.* Alexandria, VA: ASCD.

Simonson, M. et al. (2008). *Teaching and Learning at a Distance: Foundations of Distance Education.* 4th Edition. Upper Saddle River, NJ: Prentice-Hall.

Spector, J. M., & Harris, P. A. (2007). *Handbook of Research on Educational Communications and Technology.* New York: Routledge.

Stephen Heppell's Weblog: http://www.heppell.net/weblog/stephen/

Students as free agent learners: http://thejournal.com/articles/2009/04/24/students-as-free-agent-learners.aspx

Tapscott, D., & Williams, A. D. (2008). *Wikinomics: How Mass Collaboration Changes Everything.* Expanded Edition. New York: Portfolio Press.

Technology in Education: http://www.ncrel.org/sdrs/areas/te0cont.htm

Technology in Schools: What the Research Says: A 2009 Update http://www.getideas.org/sites/default/files/research/Technology_in_Schools_2009_-_What_the_Research_Says.pdf

The Brookings Institution. *Children and Electronic Media.* (2008). Paper presented at A Future of Children Event of the Princeton-Brookings Institute, Washington, DC.

The Committed Sardine by Ian Jukes: http://web.mac.com/iajukes/thecommittedsardine/BLOG/BLOG.html

The Critical Thinking Community: http://www.criticalthinking.org/

The New Untouchables: http://www.nytimes.com/2009/10/21/opinion/21friedman.html?_r=2

Time to Act: An Agenda for Advancing Adolescent Literacy for College and Career Success: http://www.carnegie.org/literacy/tta/

Top Eleven Things All Teachers Must Know About Technology (or: I promised Dean Groom I wouldn't write a top ten list; so this one goes up to eleven). http://teachpaperless.blogspot.com/2009/07/top-eleven-things-all-teachers-must.html

Trilling, B. & Fadel, C. *21st Century Skills: Learning for Life in Our Times.* Jossey Bass, 2009.

Twelve Essentials for Technology Integration: http://www.freetech4teachers.com/2009/06/twelve-essentials-for-technology.html

Using Technology Tools to Build Excellence in Teaching & Learning: http://sites.google.com/a/schoollearningcommons.info/home/using-technology-tools-to-build-excellence

Warlick, D. (2007). *Classroom Blogging.* 2nd Ed. Raleigh, NC: Lulu.com.

Warlick, D. 2¢ Worth: Teaching & Learning in the New Information Landscape. Available at http://davidwarlick.com/2cents/

Welcome to the Library. Say Goodbye to the Books: http://www.boston.com/news/local/massachusetts/articles/2009/09/04/a_library_without_the_books/

Why Schools Should Break the Web 2.0 Barrier: http://www.ciconline.org/thresholdsummer09

ARTICLES IN
TEACHER LIBRARIAN

• Biladeau, S. (2009). Technology and Diversity: Perceptions of Idaho's "Digital Natives." 36(3), 20–21.

• Davis, V. (2009). Influencing Positive Change: The Vital Behaviors to Turn Schools toward Success.
• Derry, B. (2008). Information and Technology Literacy. 36(1), 23–25.
• Endicott-Popovsky, B. (2009). Seeking a Balance: Online Safety for Our Children.
• Nevin, R. (2009). Supporting 21st Century Learning through Google Apps.
• Calamari, C. (2009). WLANS for the 21st Century Library.
• Byrne, R. (2009). The Impact of Web 2.0 on Teaching and Learning.
• Clark, B., & Stierman, J. (2009). Identify, Organize, and Retrieve Items Using Zotero.

APPRECIATION

Our appreciation goes out to the following individuals who contributed to this document: Joyce Valenza, Carol Koechlin, Sydney Cohen, April Gilbert, Terence Krista, Kathleen Riley, Susan Blair, Dana Stemig, Karen Lee, Jennifer Schwelik, and *TL* advisory board members Doug Johnson, Michele Farquharson, Erlene Bishop Killeen, Susan Ballard, and Connie Champlin.

Elizabeth "Betty" Marcoux, a part-time faculty member of the Information School, University of Washington, is a co-editor of *Teacher Librarian.* She may be reached at *b.marcoux@verizon.net.*

David V. Loertscher is coeditor of *Teacher Librarian,* author, international consultant, and professor at the School of Library and Information Science, San Jose, CA. He is also president of Hi Willow Research and Publishing and a past president of the American Association of School Librarians. He can be reached *davidlibrarian@gmail.com.*

FEATURE ARTICLE

Supporting 21st Century Learning Through Google Apps

"Google Apps has the most popular applications available as part of their cloud including a word processor as well as spreadsheet and presentation software."

ROGER NEVIN

According the 2009 *Horizon Report*, which "draws on a comprehensive body of published resources, current research and practice" there are "six emerging technologies or practices that are likely to enter mainstream use in learning-focused organizations within three adoption horizons over the next one to five years."

At Adam Scott Collegiate and Vocational Institute (CVI) in Peterborough, Canada, the Learning Commons is the school's central place for introducing new technology to students and teachers. This includes collaborative projects with teachers to develop curriculum that uses new technologies. Over the last few years podcasting as well as other online and video collaborative e-learning tools have all been implemented through the Learning Commons.

Currently, as the teacher-librarian at Adam Scott CVI, I am piloting two new projects—Google Apps and Netbooks. Both pilot projects support the two most important trends in educational technology according to the *Horizon Report*—cloud computing and portability. These projects had an immediate and significant effect on improving learning for students as well as inspiring teachers to integrate technology in the classroom in order to engage students.

CLOUD COMPUTING & GOOGLE APPS

Google Apps uses the paradigm of cloud computing. Anyone who uses common email systems such as Hotmail or Gmail already uses cloud computing. It does not matter what computer is used to access email because both the program (to run the email system) and the data used by the system (e.g., the text of messages or attachments) are located on an Internet server, which is referred to as the "cloud." The Internet server could be located anywhere in the world, but from the user's point of view, it does not matter where the "cloud" is located.

Google Apps has the most popular applications available as part of their cloud including a word processor as well as spreadsheet and presentation software. All that is needed to use Google Apps is a browser and access to the Internet.

These applications (Word, Excel, PowerPoint, etc.) allow students and teachers to create documents, share calendars, email, chat, create web pages, video, and more. It is secure as everything stays within the registered domain and cannot be accessed by people who do not have a login. It is an excellent tool to provide e-learning. It works on any computer including Macs and many Personal Digital Assistants (PDAs) such as cell phones, iPhones, and netbooks.

There are thousands of schools and millions of students around the world registered for Google Apps Education Edition. These include secondary schools, colleges, and universities including Arizona State University, Trinity College

Dublin, and Notre Dame (see www.connectingeducation.com to read case studies):

For instance, Kevin Roberts, chief information officer of Abilene Christian University, reports, "The school dumped its own e-mail program in exchange for Google Apps in 2007...it freed us up to concentrate on classroom applications."

Also, Andrew Stillman, Assistant Principal of Columbia Secondary School says, "Our technology and information systems are a huge selling point for parents [who] may otherwise have doubts [about sending students to our school]. "

After all, Google Apps is free for non-profit educational institutions, so it has the potential to save school districts significant amounts of money because Google Apps replaces most of the other software used and much of the physical infrastructure such as school and district servers. This means money can be redirected from IT into the classroom, which can have a significant effect to improve learning:

"Frantic troubleshooting by an overworked staff versus someone else fixing problems smoothly. A sliver of server space per person versus a five-gigabyte chunk. Half a million dollars versus free. That's what colleges are faced with as they decide whether to continue running their own e-mail services or outsource them to a professional service like Google Apps Education Edition" (Carnevale, 2008, p. A1).

"Our project cost one tenth of what it would if done internally" (Paul Duldig, University of Adelaide vice-president of Services and Resources—personal communication, February 2009).

TYPICAL SCENARIOS THAT ARE SOLVED WITH CLOUD COMPUTING

When students are using different computer systems and different software between school and home, anything and everything can happen.

- Jason arrives to school with his world issues major essay on a USB key. He tries to open it at school but is unable to because his word processor at home is incompatible with the school's word processor. His essay is due in 20 minutes and Jason cannot get home to print it.
- Maria creates a presentation for her law class on a Macintosh computer at home. Unfortunately all the computers at the school are PCs and she is unable to show her presentation.
- A group presentation is assigned to an English class. In one group, three students have part-time jobs, two students are on the rugby team, and one student is trying to earn her 40 community hours. They are unable to get together after school to work on the group project because of their incompatible calendars.
- You are teaching a Grade 12 history course and the final June essays are due. One of your students comes to you and says that his computer crashed and he lost all his work. He's not sure when he will be able to submit the paper.

Google Apps eliminates all these problems and provides an integrated solution where students have access to their work both at school and at home in a collaborative environment without the worry of losing data. Google Apps helps make the technology easier so students can concentrate on learning.

ADVANTAGES OF CLOUD COMPUTING

Cloud computing is popular because it has many advantages over traditional computer systems where programs are located on a computer's hard drive.

- Software is available for free and it does not have to be installed. Also programs do not have to take up hard drive space on the computer.
- Software versions are automatically updated when new features are added.
- Documents are automatically saved. No more lost documents even if the computer crashes.
- Documents can be shared in real time with other users. Students can easily collaborate for group projects and it also allows the teacher access to students documents while they are working on them.
- Documents can be published as web pages.
- Students can access Google Apps from their cell phones and any device that has access to the Internet.
- Reduces the need to print, because of access to documents at home and on PDAs. This helps the environment and saves schools money.

Another important advantage is that students are using exactly the same computer environment at home and at school. This means students only have to learn one environment and do not have to worry about using different software between school and home. Also students receive free software for home use; this is especially good for presentation and spreadsheet software, which students often do not have at home.

Google Apps provides students with unique tools such as the ability to create online surveys where spreadsheets are automatically updated with data that is collected over the Internet. They can publish any document as a web page, making students "global citizens" and giving them the ability to make global connections. They can access educational Google gadgets and even create their own.

Google also provides a top rated online calendar system. Teachers use their calendar to communicate to students important class dates and extracurricular dates (sports events, club meetings, trips, etc.) Students can also create their own calendar and mesh it with other calendars. Teachers' calendars can be posted as web pages so parents have access to class activity schedules.

Google has created a web site containing lessons and assignments on using Google Apps for different subjects and grade levels. This web site continues to grow as more schools use Google Apps and is building a community of educators who share ideas and resources. (See www.connectingeducation.com for the Google Apps assignment and resource page.)

GOOGLE APPS IMPROVES ASSESSMENT

Often students are surprised when they get an assignment back from their teachers who have given them further instructions. The student followed these instructions and submitted the work only to discover their assignment was assessed as being below standard.

Google Apps allows students to avoid the shock of receiving a low mark—after it is too late—because they can share their documents with the teacher as they are working on them. The teacher can make constructive comments right on the assignment document. This gives the student the opportunity to improve their work before they submit their final version.

Also, when students create documents, not only are they automatically saved every few minutes, but every single revision is recorded. The teacher is able to see every revision and the number of revisions, and so is an excellent way to eliminate plagiarism.

In our grade 12 English course, students are required to do all work on their major essay, including rough notes, on a Google Apps document. This document is then shared with the teacher. Typically a 3000 word essay would have anywhere from 70 to 300 revisions. If there were only one or two revisions then we assume the student copied the text from another source and pasted into it into the Google Apps document.

When students are working on a shared group assignment document, Google Apps automatically records who did what work. Students working in groups could use this to put pressure on group members who are not working. Also when the document is shared with the teacher, the teacher can see who is not pulling their weight. Since the document is shared in real time, the teacher can contact students to ask them to contribute before the assignment is due, and before it is too late.

Another advantage for collaborative documents is that they can be viewed by a teacher who may have a concern about a student. Imagine the following scenario wherein a student starts Grade 9 and is given a Google Apps login, which he keeps all the way through Grade 12. Most of the assignments the student does are on Google Apps. Many of these assignments are shared with teachers who make comments and give a score that is recorded on the assignment. A student entering Grade 10 and having trouble in his English class can immediately gain assistance from the teacher. The English teacher can go back and look at the student's Grade 9 English assignments to see the comments and the grade given by that teacher.

GOOGLE APPS AND VIDEO

Students are part of the Youtube generation and many are comfortable making videos. At my school, teachers often give their students the opportunity to create videos as part of their assignments or projects. Google Apps has a module that allows teachers and students to upload videos in the same way as Youtube. The main difference is that these videos are not available to the public. They are only available to users in the domain. Additionally, whoever uploads the video can specify exactly what users have access to view it. These videos then become stored as part of the domain and students and teachers can rate or add their comments to them (there is an option to turn this feature off). Students can also upload projects and footage of class trips while teachers also create tutorial and informational videos. Once the videos are uploaded they can be embedded into a web page.

IMPLEMENTING THE PROJECT

Google gives schools two options to register and setup Google Apps. They can either register and set up Google Apps themselves or they can go through a Google Apps partner who will charge for the service. These partners are listed on the Google Apps web site.

At Adam Scott CVI, I registered and set up Google Apps myself. The process was straightforward because Google provides very clear information and even how-to videos. To register for Google Apps, Google requires proof your school is a non-profit educational institution and approval from a school administrator. The school needs a registered web address, which will become part of the Google Apps domain. So after completing the registration for all applications—calendars, emails etc; you will use that web address.

Getting Started

When I set up Google Apps in January 2008 for my school, it took about two weeks from the beginning of the registration until I had all students on the system with logins. Currently, because of the increase in demand, the registration process can take up to six weeks.

The creation of user logins was simple. Google only requires a text file with the first name, last name, login, and password of each student and teacher. This is important because Google does not require phone numbers, addresses, or any other identifying data and so if an outsider was ever to hack into the domain they would not have access to private data. Most schools already have text files with student data that can be used to create batch files so the logins can be created automatically. The whole procedure took less than 30 minutes for around 1000 users.

Initially two classes were selected to test Google Apps—both from Grade 12 English. These classes were selected because I felt senior level students would learn Google Apps more quickly. Later when I implemented Google Apps in lower grades I found that virtually all students from Grades 9 to 12 had little problem learning how to use Google Apps.

The first teacher to use Google Apps with his class was an English teacher with "average" technical skills. The teacher had little problem learning Google Apps, even though, not surprisingly, he did not learn it as fast as most students. (See www.connectingeducation.com for videos of teachers talking about Google Apps, including Cory Pavicich, Watershed School's (Ontario) Director of Educational Technologies, who says: "Our veteran staff, almost without exception, asserts that this is the best classroom software they've ever encountered."

The Follow Through

Before students were given their Google App logins they were required to have a permission form signed by a parent or guardian. See the site www.connectingeducation.com for the permission form; any teacher can edit this form and use it for their school.

Each class was given about 20 minutes of instruction on using Google Apps. Students were also instructed *not* to use Google Apps for any personal email or chats. In other words Google Apps was for schoolwork and if they wanted to socialize, they had to use their personal email and social networking sites. Since starting the project over a year and a half ago, I have not encountered any problems with students using Google Apps inappropriately.

I surveyed students on how they liked Google Apps and whether it was helping them in their classes. Most students felt it was a great tool because they could use it at home and school and they never lost documents. I also noticed students who were only required to use Google Apps in their English class started using Google Apps in other classes including Math and Geography—even though those classes were not using the service officially. I even started to get students who were not part of the Grade 12 English classes asking for logins.

After the success with the grade 12 English classes, I introduced Google Apps to other classes from grade 9 to grade 12. Math teachers really liked Google Apps because their students were able to work on spreadsheets at home (many students did not have the software at home). In creating a web site, the Geography teachers used the Sites feature to add videos and Google Gadgets right onto the class site. Other teachers also got on board with Google Apps because their students requested it. The students were using it in their other classes, saw the benefits, and wanted all their teachers to use it.

CHALLENGES IMPLEMENTING GOOGLE APPS

There is resistance to the use of Google Apps in some educational communities because student and teacher data including all email is located on a server outside the domain of the school district. The main concern is that privacy and security may be compromised.

As of the writing of this article, I have not found any breaches concerning the security or privacy of any student or teacher, and this includes millions of users. I also contacted Google who said they know of no such case. A much larger concern is how students use their personal email and social networks where students are completely exposed to the public.

Google has responded to these concerns with a free offering of their Postini Services, a product specifically designed for Internet security, with the Google Apps Education Edition. Postini can be used to ensure email is only sent to addresses within the domain and only approved email addresses outside the domain can be accessed. It also has filters to catch students who use inappropriate language or who may engage in cyberbullying. And, it keeps emails that have been deleted for auditing. Also the virus and spam checker is updated on a daily basis.

CONCLUSION

Google Apps has significantly improved the way students and teachers work at Adam Scott CVI, providing a common collaborative system that virtually supplies all the applications and communication tools needed, under one platform, and at no extra cost. It makes using technology easier as the computer environment is the same at school and at home. Additionally, because Google Apps eliminates or limits the need for printing and photocopying, it is both cost effective and easier on the environment.

REFERENCES

Carnevale, Dan. (2008). Colleges get out of e-mail business. *Chronicle of Higher Education*, (54) 18.

Johnson, L., Levine, A., & Smith, R. (2009). *The 2009 Horizon Report*. Austin, Texas: The New Media Consortium. Retrieved from http://www.nmc.org/pdf/2009-Horizon-Report.pdf.

Roger Nevin has taught for over 26 years at the secondary, college, and university level, and is currently the teacher-librarian at Adam Scott C.V.I. He has presented over 26 times on implementing new technology in the classroom including seven conference presentations, which includes 2 national conferences. He is author of *New Approaches For Learning 21st Century Skills* and is a cofounder of **boysread.com**, a non-profit web site supporting male literacy, as well as **connectingeducation.com**, another non-profit web site that provides resources for educators who want to connect education with young technology users. He may be reached at *connectingeducation@gmail.com*.

Note: To view Google Apps at work and for all links, podcasts, and videos mentioned in this article, visit **connectingeducation.com**, a non profit educational web site.

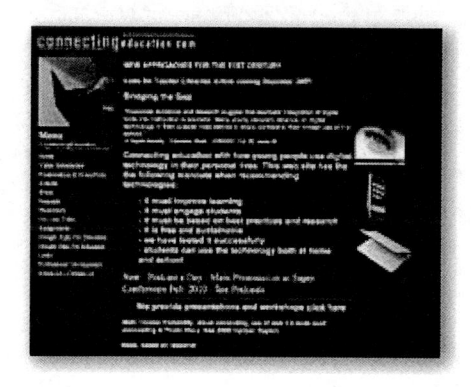

TL EXTRA

The Effect of Web 2.0 on Teaching and Learning

"The fact of the matter is that the longer schools wait to use technology in their classrooms, the further behind their global peers students will become."

RICHARD BYRNE

We are nearly a full decade into the 21st Century and unfortunately many schools are still using the term "21st Century teaching" as a far-off, futuristic idea.

The fact of the matter is that the longer schools wait to use technology in their classrooms, the further behind their global peers students will become.

In many cases the resistance from school administrators comes in the form of, "we don't have the money." At the same time resistance comes from some teachers who claim, "I've done it this way for X number of years, and my students have done just fine." A third claim that school administrators and educators occasionally make is, "technology won't improve student performance." There is an element of truth in those claims, but that does not excuse not making a full-fledged effort to integrate technology into K-12 classroom instruction. With some research, creativity, and professional development, any school can stop talking about becoming a 21st Century school and confidently become a 21st Century School.

MERGING TECHNOLOGY WITH TEACHING

Can a teacher be confident that if students embrace all the wonderful tools, they will go beyond minimal test scores and on into the world of deep understanding, critical thinking, and creative thinking, and gain the technological expertise they need to excel globally? Absolutely yes!

Three years ago I was not aware of 10% of the free educational resources available on the Internet that I am aware of today. The acquisition of new knowledge began when I was given the opportunity to pilot the use of classroom laptop computers. At the time I was teaching a social studies survey course for high school freshmen.

One element of the course was an exploration of the policy-making process of the United States Congress. I had taught the course before and students had to complete two assignments for the unit: a flow chart and a mock debate related to a current policy issue. Suddenly, having laptops in the classroom gave me an opportunity to try something new. For instance, a quick Google search for "teaching about Congress" brought me to The Center on Congress at Indiana State University, http://congress.indiana.edu/index.php. One of the featured activities on the site was an interactive activity called *How a Member Decides to Vote*, http://congress.indiana.edu/learn_about/launcher.htm. This free, web-based activity provided students with the opportunity to take on the persona of a congressman or congresswoman. Throughout the activity the students would receive calls from constituents, talk with special interest groups, attend committee meetings, and attend votes in Congress. The calls from constituents would come in at random intervals throughout the activity. This activity gave students the opportunity to experience the chart rather than just study and create flow charts. At the end of the unit of study on the policy making process, that year's students had a better understanding of the process than did the students of prior years.

This story is one of personal success in one classroom, but more important, it began a grassroots movement of teachers interested in learning about the free technology resources available for classroom use. For the most part, the teachers' interest was piqued by hearing students talking about what they were doing in my class. It is no secret that students who are excited about coming to class and are engaged in class will perform better. Technology in and of itself does not create engaged students, but using web applications that allow students to create new content does engage students in learning.

So can a teacher be sure of that students will perform better, delve deeper, and become more creative because they have integrated technology into their classroom? The answer is a resounding yes if you believe improved student engagement improves learning.

TECHNOLOGY, IT IS NOT JUST A BUNCH OF PLAY TOOLS

Technology in and of itself will not create more engaged students or better students. However, well-chosen technology resources infused into classroom instruction can create more engaged and better students. Given the choice of having students create content or simply absorb content through reading, listening, or viewing, I will choose creating content every time.

Part of my teaching load last year, as it has been in most years, was a United States history course for special education students. The typical student in the course is seventeen or eighteen years old with a reading ability of an average third or fourth grade student.

Last year, I was able to have almost uninterrupted access to a computer lab adjacent to where I taught the group. Just as with using the laptops for my freshman course, having access to the Internet for my special education group enabled me to provide my students with more engaging learning opportunities. Before getting Internet access for these students, creating differentiated and engaging lessons for the class was the most challenging part of teaching the course.

Before the Internet, options for creating content in the United States history course were generally limited to writing, drawing, and occasional acting. Not only were the forms of creating content limited, the audience for the content my students created was also limited. Consistent access to computers and the Internet gave me the ability to offer my students a number of new ways to create content and to share that content with a larger audience using the immensely engaging Animoto, http://animoto.com.

What makes Animoto so engaging is that it allows students to create very professional-looking music videos on any topic, without the need for special technical skills. To create videos, students need to locate images, upload those images to Animoto, insert text, and select music tracks. Animoto then blends the elements together and produces a high-quality video.

The students in the history class made videos as part of a summative digital portfolio for a marking period (semester in the United States). They created a series of Animoto videos, about different topics as we progressed through our study of the 20th Century. Each time we completed an era or decade, the students created a video about that topic. The students enjoyed making the videos so much they begged to make more everyday. Therefore, I was able to leverage my students' excitement to keep them focused on academic learning during class.

CAPITALIZE ON MYSPACE AND FACEBOOK SKILLS

A couple of years ago while preparing to give a presentation about the educational uses of blogs and wikis, I came up with the phrase, "capitalize on your students' MySpace skills." Ask your students how many of them have a MySpace or Facebook account and you are likely to see every hand raised. It follows that if your students can manage a social networking profile, they can use a wiki or a blog.

Adding content to a wiki or a blog takes no more skill than modifying a MySpace profile. The trouble is, your students may not know this and if they do know it, they may not know what to add. That is where you come into the picture. Your role as the teacher is to give purpose to the MySpace skills your students have developed. A simple project that I developed a few years ago for one class was to build a wiki about Africa. At the time, none of my students had added content to a wiki, but they had added content to MySpace. Because they already had email accounts and knew how to create profiles, after a short lesson on how to use a wiki, my students were off and running. For the project each student was given an African country to research. Each student then added the information they found to their own page on the wiki. By the end of the first day of the project, students were competing against each other to create the best wiki page; the public nature of the wiki compelled them to put extra effort into the project.

If you have never created a wiki or blog, do not worry, there are many excellent tutorials online including on YouTube. In many cases your students will figure out a wiki or blog faster than you will. If you are in that position, the best thing to do is embrace it as a learning experience to share with your students.

BEYOND CUT AND PASTE

With today's endless supply of online content, cutting and pasting is easy to do, but does not prove that the student has learned anything. The students who created wikis about Africa put extra effort into the aesthetics of their pages, but also into the depth of content on their pages.

Prior to having access to the Internet during classes, students were limited to content in the books in the school library. However, the option to click from link to link to track sources of the information displayed by various web sites, gave my students access to content they would not otherwise be able to access. In their quest to show how much they learned about their assigned country, my students dove far deeper than standard reference sources in both print and online.

GETTING ACCESS TO THE INTERNET

At least once per week, I receive emails from readers of my blog lamenting that a resource I have shared is blocked in their schools.

I have been fortunate to work in a school district, Oxford Hills School District in Maine (MSAD#17), that prides itself on making research-based decisions. Time and time again, research like that found in the MacArthur Foundation's Digital Youth study (2008) indicates that open access to the Internet is educationally valuable. In the case of my school district, that research is the basis for our open access policy. Unfortunately, many school districts across the country block access to valuable online resources because they are fearful of students accessing questionable content and/or are fearful of students' misconduct online. The alternative to censoring is to educate students about responsible online behavior. If the role of the school is to educate the whole student, teaching students how to behave online has to be a key component to the education of the whole student.

GAUGING THE EFFECT OF TEACHING WITH TECHNOLOGY

In today's results-oriented school culture, being able to measure the effect technology in the classroom has on student learning is important to many teachers, librarians, and school administrators. This is easier to do in some content areas and grade levels than it is in others. I have developed a checklist of benchmarks to gauge the benefits to students and teachers of teaching with technology:

Efficiency (students work smarter and so do teachers)

- After proper instruction, students access more and deeper information.
- After proper instruction, students can create a document, video, podcast, or presentation that demonstrates a deeper understanding of their content area(s).
- Teachers bring more current and relevant teaching materials into their lesson plans.
- Hyperlinked writing makes it easier for teachers and students to verify the information presented in student work.
- Students bring more information into classroom discussions, written work, and multi-media presentations.

Motivation to Learn

- Students come to class suggesting or asking to develop a learning project.
- Students access and use information to challenge each other's statements.

Deep Understanding

- Students create content that refers to and builds upon references that are more in-depth than those found in classroom textbooks.
- Teachers use the Internet to stay current on best practices and to develop lessons that provide students with opportunities for deeper learning.

Learning How to Learn

- Students not only answer questions posed to them, but create their own questions based upon the wealth of information they are able to access.
- By following links online, students develop the habit of verifying information and locating deeper information in the style of bibliography-chasing used by students of previous generations.

One practice I have started using in my classroom that has certainly lead to deeper understanding is having a "Google jockey" for each class. This practice, which I learned about in Curtis Bonk's book *The World Is Open* (2009) assigns one student in the classroom to be in charge of Googling any new terms or questions that arise during a lecture or discussion. Depending on the day, the Google jockey may also be charged with finding information to add to the discussion in the class. In the short time that I have used the Google jockey practice, I can confidently say the students are accessing more information and learning more about each topic than they could if they were relying solely on print materials and my knowledge.

BEFORE AND AFTER, ONE TEACHER'S EXPERIENCE

As much as I enjoy discovering new content and the latest web tools, none of it would matter if it did not have a positive effect on my students' education. After teaching without using technology in my classroom and then teaching with it, I know that web-based resources have improved my students' education. A review of the grade books from the last three years of teaching the United States History course for special education students reveals that the average grade has increased over that span of time. Over the same span my technology use with that course has also increased.

If you are reading this article and are ready to charge onto the Internet in search of new resources for your students, it is important to keep the following in mind. First, keep abreast of available technology and develop new skills. To help you along the way, check out "Free Technology for Teachers" (http://freetech4teachers.com), Larry Ferlazzo's "Websites of the Day" (http://larryferlazzo.edublogs.org/), and Wesley Fryer's "Moving at the Speed of Creativity" (http://www.speedofcreativity.org/). Second, and most important, remember that technology alone will not improve students' education. However, technology appropriately chosen and appropriately integrated into a lesson can improve your students' engagement and in turn improve your students' education.

REFERENCES

Bonk, C. (2009). *The world is open: How web technology is revolutionizing education*. San Francisco: Jossey-Bass.

Ito, Mizuko, et al. (2008). Living and learning with new media: Summary of findings from the Digital Youth Project. Chicago: The John D. and Catherine T. MacArthur Foundation. Retrieved September 30, 2009 from http://digitalyouth.ischool.berkeley.edu/report.

Richard Byrne, teacher of United States History and Contemporary World History at Oxford Hills Comprehensive High School in South Paris, ME, is also author of the blog, Free Technology for Teachers, www.freetech4teachers.com, which features reviews and ideas for using free web-based resources in education. In 2008 the web site was awarded the Edublogs Award for Best Resource Sharing Blog. He may be contacted at *richardbyrne@freetech4teachers.com*.

TLEXTRA

WLANS for the 21st Century Library

"A wireless network that fails to provide secure, predictable, and reliable access will frustrate users and library staff."

CAL CALAMARI

As educational and research needs have changed, libraries have changed as well. They must meet ever-increasing demand for access to online media, subscriptions to archives, video, audio, and other content.

The way a user/patron accesses this information has also changed. Gone are the days of a few hardwired desktops or computer carts. While libraries still need to support people without computers, many users bring their own wireless laptops, Netbooks, Tablet PCs, E-book readers, or smartphones. Whether using these for school studies, teaching, researching, or pleasure, users want to access information without switching between two computers.

To alleviate the expense, headaches, and administration of wired connections, libraries have turned to Wireless Local Area Networks (WLAN) for primary network access. Going wireless eliminates the need to have fixed location PCs, the headaches of too few wired ports, and troubleshooting broken cables. By switching to WLAN, the library evolves to a 'learning commons'—creating a collaborative learning environment and providing users easy access.

The learning commons creates a flexible environment where individuals, small groups, and large groups can collaborate, accessing information simultaneously with any wireless device while sharing ideas and problem solving. A learning commons encompasses both in-building and the surrounding campus, extending the library beyond the four walls to courtyards and other areas around the library. Users can connect and research wherever they are.

The library's WLAN becomes a critical service delivery medium for users and guests. A wireless network that fails to provide secure, predictable, and reliable access will frustrate users and library staff. The WLAN must provide strong authentication for controlled network access, high quality service to support various media types and devices, and be capable of scaling to dense user environments with hundreds of users at any one time.

SYMPTOMS OF WIRELESS CONNECTIVITY PROBLEMS

Early WLANs were designed as networks of convenience for users with minimal network requirements. They were limited to a few users and the applications were not time critical, typically represented by web surfing and email. However, as WLANs and Wi-Fi devices have become pervasive and applications have become more demanding, connectivity issues have increased. Disconnected sessions, slow performance as more users log in, and poor user experience with video and audio have led to an increase in user complaints and lower expectations by librarians on what they can expect from a wireless network.

UNDERSTANDING WIRELESS LAN FUNDAMENTALS

WLANs are extensions of the Local Area Network (LAN), which came before wireless was based on wired Ethernet technology. Wireless access points act as smaller versions of cell phone transmission towers, distributed to provide coverage where it is needed. Wireless clients such as laptops communicate with the wireless access points (APs).

SHARED MEDIUM

Wireless is a shared medium, meaning all devices share the same radio frequencies (RF) in the unlicensed radio frequency bands of 2.4 Giga Hertz and 5 Giga Hertz. Yes, these are the same bands for home telephones and microwaves! The fact that all the devices share the same air space without coordination creates a problem.

If two people using walkie talkies or CB radios 'key' the microphone at the same time, the signals will cancel each other out and no one gets through. Other users on the same channel will have their transmissions affected as well; they will hear a squeal as the two radios keyed will be transmitting on top of each other. These uncoordinated transmissions reduce available 'air time' for everyone on that channel, resulting in lower performance. Typically CB users will use terms like '10-4' or 'roger' as verbal protocols to help coordinate transmissions and eliminate keying on each other. This coordination improves transmission quality, prevents retries, and provides a better user experience.

WLANs use radios and can have the same interference issues if the radio transmissions are not coordinated. Imagine a library filled with 100-200 students all trying to use the wireless LAN at the same time without coordinating their transmissions. There would be a lot of squealing!

This congestion wastes air time and causes poor performance. If there are only a few users, performance might be acceptable, but as more users sign onto the network, performance will degrade, leading to users experiencing network disconnects and poor performance.

ELIMINATE CONGESTION

The best way to address the problem of congestion is to coordinate air space in a manner similar to the way Air Traffic Control manages planes at airports. Since planes share air space and runways (access points), Air Traffic Control queues planes (clients) on the ground and in the air, prioritizing them and scheduling when they will land and which runway they will use to land or take off. This eliminates congestion and increases airport throughput.

Third generation and earlier WLANs are considered microcell architectures because they work in the same way as an old, analog cell phone system. They do not coordinate airtime transmissions and cause airtime contention between clients and access points. This contention causes delays and only gets worse as more clients are on the network.

Fourth generation WLAN architecture eliminates the contention by using a similar approach to Air Traffic Control, queuing clients according to their application needs, and scheduling when and to which access point they will transmit. This technique eliminates congestion, provides predictable connections, and increases the number of clients that can be supported.

REAL TIME APPLICATIONS: VOICE, AUDIO, AND VIDEO

In traditional wireless LANs, otherwise known as microcell, transmissions are not scheduled—other clients can be transmitting at the same time. These air time collisions will cause retries until the transmission can be complete, resulting in degrading performance, which further affects the predictability of transmissions.

Real time applications such as voice, audio, and video are highly sensitive to these variations in the system. User's ears and eyes are very perceptive to these variations and can discern frame freezing, jittery video, and dropped packets in voice calls, all of which lead to lower quality and poor user experience.

When wired Ethernet first came onto the market it was a shared medium, based on devices called hubs that broadcast every computer's traffic to every other computer—just like older wireless LANs. However, Ethernet evolved to switching, which means the wires leading to each computer are separated so each gets its own private connection with port level control, security, and quality of service (QoS) for each device connected to the switch. By doing so, real time voice and video applications will receive the appropriate level of QoS, while voice running on the adjacent port will receive a different level of QoS to meet the requirements of voice.

Switch-based wireless LANs offer virtual ports to each client using Air Traffic Control, providing a similar level of control over each wireless device and automatically applying appropriate levels of QoS for the application running on those devices via application aware QoS. Virtual Port provides clients isolation, allowing each client to operate at maximum potential without affecting nearby clients, while ensuring the application provided the appropriate QoS for the best user experience.

MIXED CLIENTS

As mentioned earlier, there will be many users with various WiFi™ device types. These devices can have a range of WiFi radios such as 802.11b, 802.11g, or 802.11n. Each device has a different performance capability, ranging from 1 megabit/s to 300 megabit/s. The problem with traditional hub-based WLANs is that in mixed client environments they will reduce the protocol rate to the lowest common denominator. If there are users with 802.11n running at 300 and someone comes in with an iPod running 802.11b, everyone's performance will be reduced to 11b's speed of 11 megabits. This is not good if the user paid for a new 802.11n card expecting great performance.

The way you really want the system to

ABOUT MERU NETWORKS

Founded in 2002, Meru Networks develops and markets wireless infrastructure solutions that enable the All-Wireless Enterprise. Its innovations deliver pervasive, wireless service fidelity for business-critical applications to major Fortune 500 enterprises, universities, healthcare organizations, and local, state, and federal government agencies. Meru's Air Traffic Control technology brings the benefits of the cellular world to the WLAN environment, and its WLAN System is the only solution on the market that delivers predictable bandwidth and over-the-air quality of service with the reliability, scalability, and security necessary to deliver converged voice and data services over a single WLAN infrastructure. For more information visit www.merunetworks.com.

work is to automatically assign each device airtime, providing a fixed amount of time that they can transmit at their full protocol rate without being affected by other protocols. This will allow 11b, 11g, and 11n devices to operate in the same RF band, but each achieving the performance the user expects.

WIRELESS LAN REQUIREMENTS FOR 21ST CENTURY LIBRARIES

1. Full strength radio transmissions for difficult environments.

Library environments represent a challenge for radio frequencies (RF) due to the dense collection of books which absorb moisture and RF signals. Libraries may also have large open spaces, high ceilings, and thick walls. Some libraries may reside in historical buildings requiring minimal wiring.

Requirement: *WLAN access points should be deployed at full RF power allowing overlapped RF coverage using a single RF channel, ensuring greatest signal transmission, penetration, and receptivity with no dropped connections or disconnects of user device.* Channel layering, allowing pervasive RF channels to be stacked in the same area, should be supported to provide additional capacity or redundancy. Operating at full RF power will reduce the number of access points required, along with associated cabling, which is particularly important in historic buildings. The Wireless LAN system should have a simple deployment methodology to minimize pre-deployment expenses, eliminating costly RF surveys, and minimize day-to-day management.

2. Predictable and reliable service delivery.

Libraries are service providers, providing access to information for users and guests. Information resides in the form of books, research articles, news clips, videos, and audio, representing many media types the system must accommodate. The wireless LAN is the primary conduit for delivery of that service. If the library cannot provide those services in a predictable and reliable manner, users will lose valuable research time, become frustrated and log complaints. This further burdens librarians, their staff and IT departments, affecting their productivity.

Requirement: *The WLAN must provide predictable and reliable service delivery of all media types; voice, video, and data, to dense user environments, supporting various WiFi devices without performance degradation.*

3. Support for dense user environments.

Libraries are typified by large numbers of guests using the resources around the same time. One of the most difficult challenges for wireless systems is to support dense user environments when there could be hundreds of users trying to access the network at the same time. If users cannot get connected or have their connections dropped in the middle of a session, they will have to try reconnecting and restarting their sessions, potentially losing content.

Requirement: *The wireless LAN system needs to support 50 to 100 users per access point while under load, supporting multiple device types, resulting in no dropped connections.*

4. Support for all devices without degraded performance.

Since users will be using their own wireless laptops, Netbooks, iPhones, Blackberries, etc., each device will have its own WiFi adaptor and each will support a specific IEEE 802.11 standard, such as 802.11b, 11g, 11a, or the latest standard of 11n. Each of these standards operates at a different performance level, ranging from 1 Mbps (mega bits per second) to 300 Mbps. Wireless LAN systems will default to the lowest performing device when a mixed set of devices are operating together, reducing the performance of all the other devices associated to that access point to the lowest common denominator. In an environment such as a library, one user with an old 802.11b laptop will penalize all users, affecting all other device performance. The users may think the wireless LAN is at fault, while it is the old device using the same RF space.

Requirement: *The wireless LAN system must support all Wi-Fi certified devices that may have 802.11 a/b/g/n. The wireless LAN must also provide air time fairness so that each client operates at its maximum rate without being affected by other clients.*

5. Secure Network Access.

Though libraries are generally viewed as available to all students, teachers, and possibly the public, libraries often need to authenticate users and limit their access to network services or resources and sometimes limit access to some web content based on user role or location. Allowing any and everyone open access to the network without encryption and authentication could compromise user privacy and integrity of the network and its content, which would be detrimental to all.

Requirement: *The WLAN system must support all WiFi certified encryption protocols (most broadly used today are WPA and WPA2 pre-shared Key) to protect all wireless transmissions.* The WLAN system must also provide guest portal control, which will provide users a login screen to enable user authentication and provide user classification and Per User Firewalls, limiting network resource access.

AND FINALLY...

Our advice to teacher-librarians is to become informed enough about these networking issues so you can join the conversation at school and district levels armed with the knowledge of how the user is reacting to the various networking issues and problems and also with the familiarity of the latest in networking issues so that your seat at the table is a powerful one. You are the advocate for the users from the front line and our responsibility as network creators and builders is to install systems that are reliable, fast, friendly, and that will support the number of users that need to be working simultaneously. You know when the networks are "right" because they become transparent, reliable, and available anywhere, at any time, and on any preferred device. Keep pushing until that happens.

Cal Calamari is Director of Solutions Marketing with Meru Networks, a company in Sunnyvale, CA, that develops and markets wireless infrastructure solutions that enable the All-Wireless Enterprise. He has more than 25 years of high tech as a designer and managing product lines including servers, semi-conductors and networking products. He may be contacted at *ccalamari@merunetworks.com.*

FEATURE ARTICLE

the impact of Facebook on our students

DEMONIZE IT OR EXTOL ITS ADMISSIONS AND ALUMNI-NETWORK VIRTUES, THE USE OF FACEBOOK IN OUR SCHOOLS IS LIKELY TO ELICIT STRONG OPINIONS. ONE THING IS FOR CERTAIN, THE USE OF FACEBOOK REPEATEDLY COMES UP IN DISCUSSIONS ABOUT INTERNET SAFETY, AGE-APPROPRIATE EXPOSURE, AND STUDENT ONLINE BEHAVIOR. THOUGH MANY SCHOOLS HAVE DIFFERENT POLICIES FOR USING OR ACCESSING FACEBOOK, WE SHARE MANY OF THE SAME CONCERNS.

Through our Internet safety organization, ChildrenOnline.org, we have surveyed the Internet behavior of thousands of children and teens in Independent Schools. We have learned a great deal about their use of Facebook and the inherent issues they face, as well as their schools, because Facebook is one of the two most popular web sites for students across grades 4–12. (The other site is YouTube.) We would like to summarize our shared concerns and address the issues that impact our students, and our communities.

Note: Though this article targets Facebook specifically because of its popularity, the article also applies to the many other social networks our students frequent. They include YouTube, MySpace, Hi5, Friendster, Xanga, DeviantArt, and others.

1. FOR THOSE SCHOOLS THAT ALLOW IT, THE USE OF FACEBOOK IN OUR COMMUNITIES CAN TAKE AN INORDINATE AMOUNT OF INTERNET BANDWIDTH.

And for those schools that allow access to Facebook, how do we reconcile our concerns that younger and younger children are using this adult social network? Four years ago it was rare to learn of a child under 7th grade with an account. Last fall, for the first time, 4th graders began reporting to us that they had Facebook accounts. We now estimate that about 60–70% of 7th graders have accounts and the number is higher for 8th graders. These children are too young to be using Facebook or other adult social networks for the reasons detailed below.

2. USING FACEBOOK TAKES TIME. OFTEN, A LOT OF TIME!

The greatest motivating factor for children to use technology in grades 7 and up is to connect to others; to socialize. Their irresistible need to connect with their peers, coupled with the development of 24/7 accessible technologies, can make the use of sites like Facebook all-consuming. We have concerns for children and teens today growing up in a world where they are wired 24/7 without a break. For many of our kids there is little or no "down time." Some have difficulty disengaging from their social life. For some, it even raises their anxiety level to be without their cell phones for a few hours! We do not believe this is healthy for them.

by doug fodeman and marje monroe

3. TO STUDENTS USING FACEBOOK, THERE IS A FALSE SENSE OF PRIVACY.

Couple this false sense of privacy with the feeling of anonymity and lack of social responsibility that often develops from using text-centered telecommunications, and we see that many students post embarrassing, humiliating, denigrating and hurtful content in both text, photos, and videos. We need to teach them that NOTHING IS PRIVATE online, especially their social networks. We need to show them examples of the serious consequences that have occurred to those whose egregious online behavior has been made public. Students have been expelled from high schools and colleges. Students have been denied acceptances to intern programs, admission to independent high schools, colleges, and jobs at summer camps. Students and their families have been sued for slander and defamation of character. Students and their parents have been arrested. All because of the content they have posted in their "private" social network accounts. People are trolling their accounts. Hackers, scammers, reporters, police, high school and college admissions officers, employers, parents, and summer camp directors.... Adults ARE looking and the kids do not get it! Also, they do not realize the instant they post something to Facebook (or MySpace or YouTube, etc.), they have just lost control and ownership of that content. Try reviewing the privacy rights of Facebook with your middle and high school students. It is quite an eye opener!

In fall 2007, Dr. Nora Barnes, Director for the Center of Marketing Research at University of Massachusetts Dartmouth, published a study that showed more than 20% of colleges and universities search social networks for their admissions candidates. Do you think that percentage will decrease, increase, or remain unchanged in the coming years? Ask your high school students that question!

Students often ask us, "How can anyone possibly get into my private Facebook pages?" Here are the most common methods and a link to a sample article about each:

a) Security and software flaws are exposed. Software is hacked.
—*Tech & Learning Advisor Blog*. "My Facebook account was hacked!" by Cheryl Oakes (http://www.techlearning.com/blogs.aspx?id=15098).

b) Accounts are phished when users are tricked into clicking an email or IM link taking them to fake login pages. Once phished, scammers use various applications to suck out personal information from a user's entire network of friends. Scammers try using the phished information, including the login password, to access banks and credit card accounts because they know that most people have one password for all their accounts. They also target teens' Facebook accounts because they have learned that a small percent of their parents use combinations of their children's names and birthdays as passwords to their financial and credit card accounts.
—*Tech Crunch*. "Phishing scam targeting

"The greatest motivating factor for children to use technology in grades 7 and up is to connect to other; to socialize."

"We anticipate that this year's ITL committee will use the institute as a springboard for determining goals, objectives, and school-based professional development throughout the 2008-2009 school year."

"We need to teach them that NOTHING IS PRIVATE online, especially their social networks."

"We need to teach our students that "Free" usually has a price when it comes to the Internet."

"We need to help our students become more media-savvy, to understand the value of personal information, and how to protect it."

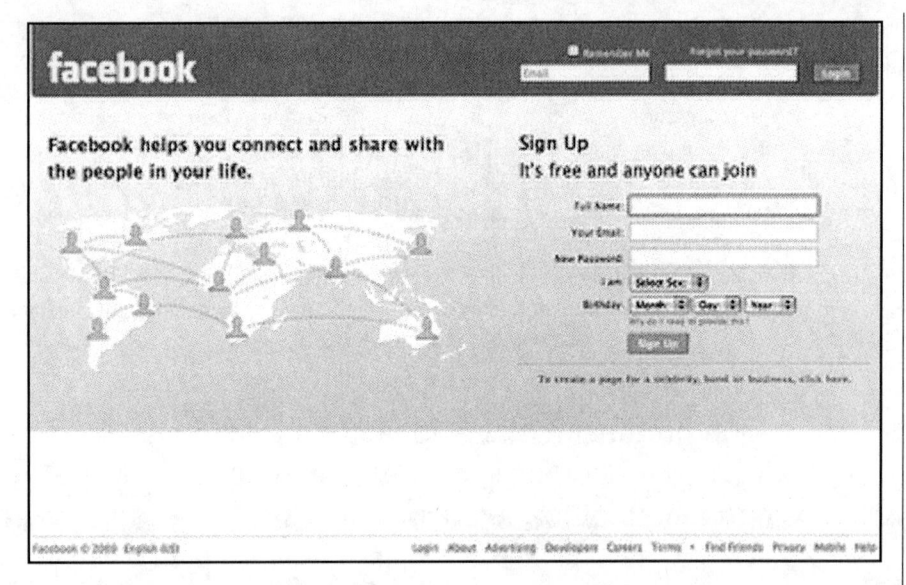

Facebook users" by Duncan Riley (http://www.techcrunch.com/2008/03/26/phishing-scam-targeting-facebook-users/).

—*Wired.* "Fraudsters target Facebook with phishing scam" by Ryan Singel (http://www.wired.com/politics/security/news/2008/01/facebook_phish).

—*The Next Web.* "Facebook under massive phishing attack from China" by Steven Carrol (http://thenextweb.com/2008/08/10/facebook-under-massive-phishing-attack-from-china/).

c) Perhaps the most common reason that teens' private information is exposed is because they are easily tricked into accepting friend requests from strangers. Though there is not a lot of research available on this point, some research and informal studies suggest that teens allow into their Facebook networks 44%–87% of the strangers that knock on their door. This trick is best described as the "wolf in sheep's clothing." Many kids, especially girls, have a difficult time saying "no" to a "friend" request.

—*C-net News.* "Facebook users pretty willing to add strangers as 'friends'" by Caroline McCarthy (http://news.cnet.com/8301-10784_3-9759401-7.html).

d) Students' passwords are easily guessed or hacked with readily available "cracking" software. We've met 5th graders who have demonstrated knowledge of using hacking tools such as password crackers. There are numerous examples of kids' accounts being hacked simply because someone guessed or figured out their password. Last September Gov. Sarah Palin's personal e-mail account was broken into when the hacker figured out that her password was a combination of her zip code and birth date.

—*CNN Politics.com.* "Agents search apartment in Palin e-mail investigation" from Terry Frieden (http://www.cnn.com/2008/POLITICS/09/22/palin.email.probe/index.html).

Note: Police, and other investigative authorities such as the FBI, can have access to "private" Facebook pages. Also, we strongly suspect that Facebook itself sells access to information posted on private pages to third party marketers willing to pay the fees. At least, that was what one former employee in the social network industry who wished to remain anonymous described to us.

4. THERE ARE THOUSANDS OF SCAMS TARGETING TEENS IN THEIR SOCIAL NETWORKS, ESPECIALLY FACEBOOK AND MYSPACE.

These communities are predicated on a certain level of trust. Our students, though very knowledgeable about using technology, are often naive and easily manipulated (though they would hate to think so). A simple example is a scam that hit Facebook users late last fall. Many teens had their accounts phished and the phishers sent out posts from those accounts to their friends that said "OMG! There are some photos of you on this web site," along with a link to the web site. The web site showed hazy photos in the background that were hard to make out and appeared to be somewhat pornographic. A popup told the visitor they would have to register for an account in order to view photos on the site. We're certain that many kids were tricked into revealing a lot of personal information in this scam. In another scam that targeted MySpace in the last couple of years, more than 14,000 users were tricked by fake MySpace pages into visiting music web sites to purchase music for $2-3 per album. Instead of getting music, the site charged their credit cards $300-600. Kids are easily fooled. They want to believe what is said to them, especially when it appears that others believe. Scammers use this trick against them by creating thousands of fake pages on social networks that talk about bogus web sites to buy stuff, products that do not work (e.g. herbal medicines) and cool pages that only result in drive-by spyware downloads.

5. SPYWARE AND ADWARE INSTALLATIONS ARE VERY SERIOUS CONCERNS.

Those of us with PCs running Windows OS in our schools already devote a great deal of time, money, and other resources to these threats. Giving kids access to social networks in our school environments greatly exacerbates these threats. We need to teach our students that "Free" usually has a price when it comes to the Internet. We need to teach them how to try to determine if software, such as a Facebook Add-on is likely a disguised piece of malware. (Much of it is!)

—*PC Advisor.* "Warning over malicious Facebook wall videos: Hackers target users with fake Flash download" by Juan Carlos Perez (http://www.pcadvisor.co.uk/news/index.cfm?newsid=102800).

—*Sophos.* "Facebook users struck by new 'court jester' malware attack posts on your Facebook wall may lead to Trojan horse infection" (http://sophos.com/pressoffice/news/articles/2008/08/facebook.html).

—*The Register.* "Link spammers go on social networking rampage" by John Leyden (http://www.theregister.co.uk/2008/04/02/facebook_spam/).

Note: "Mac owners" are not completely

off the hook. Last June, the first three spyware apps were discovered against the Mac OS and late last fall there was evidence of hijack-ware successfully targeting Firefox on a Mac. In April, 2009, the first Mac Botnet, in which scammers take control of Apple Mac computers to send out spam, was reported.

6. WE NEED TO ACKNOWLEDGE THAT SCREENS ACT AS A MORAL DISCONNECT FOR MANY OF OUR STUDENTS.

Every day online there are thousands of kids who say mean and hurtful things because they can. They are increasingly living their social lives in a world without caring, loving adults watching out for them, without expectations for their behavior, and without boundaries. Research shows that children grow up healthiest in a world with love, communication, structure, and boundaries. These qualities hardly exist online for our children/teens. Instead, harassing language is normalized, the sexualization of girls/women is common-place, and the lack of supervision creates an "anything goes" wild-wild-west. Here is a simple case in point. Would Texas Longhorn lineman, Buck Burnette, have said the same thing about President Obama if handed a microphone at a school assembly in front of hundreds of students? Would he have written his posted statement on a large poster and held it up in downtown Houston for a few hours? We doubt it.

—*Fanhouse*. "Texas' Buck Burnette learns why racist Obama Facebook updates are dumb" by Will Brinson (http://ncaafootball.fanhouse.com/2008/11/06/texas-c-buck-burnette-learns-why-racist-obama-facebook-updates-a/).

Students need to learn to be nice and kind to others online. They need to be respectful and thoughtful about what they say and how they act online, just as in real life. We need to do a better job of teaching them that disengaging from social responsibility while using telecommunications is not acceptable behavior.

7. STUDENTS HAVE VERY LITTLE KNOWLEDGE ABOUT HOW MUCH THEY ARE BEING MARKETED TO; HOW THEIR PURCHASING DECISIONS AND ATTITUDES ARE BEING MANIPULATED; HOW THEIR PERSONAL INFORMATION IS USED, AND EVEN HOW VALUABLE THAT PERSONAL INFORMATION IS.

Most do not understand the damage that can come from identity theft and impersonation. They are heavily targeted on Facebook and their data is heavily "scrubbed" and used. Facebook's announcement about Beacon in November 2007 brought such a huge negative assault from users that Mark Zuckerberg had to back-step and tell users that they were automatically opted OUT, rather than IN, as planned. Most users saw Beacon as a privacy nightmare. We need to help our students become more media-savvy, to understand the value of personal information, and how to protect it.

—*Gigaom.* "Is Facebook Beacon a privacy nightmare?" by Om Malik (http://gigaom.com/2007/11/06/facebook-beacon-privacy-issues/).

—*C-Net News.* "Facebook's Zuckerberg apologizes, allows users to turn off Beacon" by Amy Tiemann (http://news.cnet.com/8301-13507_3-9829401-18.html).

—*PCMag.com.* "Facebook's Zuckerberg apologizes for ads debacle" by Chloe Albanesius (http://www.pcmag.com/article2/0,2817,2228622,00.asp).

8. OUR RESEARCH SHOWS THAT CHILDREN AND TEENS ARE INCREASINGLY USING TELECOMMUNICATIONS TECHNOLOGIES, INCLUDING FACEBOOK, TO AVOID DIFFICULT FACE-TO-FACE CONVERSATIONS.

For example, it saddens us to hear 16-year olds say they would rather break up with their girlfriend/boyfriend by texting, IM-ing, or posting on their Facebook wall than tell them in person (or over the phone). When asked why, they'll tell you "because it's easier." We believe this avoidance will have increasingly negative ramifications on their communication skills throughout life.

9. MORE AND MORE, CHILDREN ARE TURNING TO MAKING FRIENDSHIPS AND BUILDING RELATIONSHIPS ONLINE.

This includes the use of Facebook. Socialization skills in children are best learned in real life. Children are far too inexperienced to use telecommunications tools to make friends and build relationships in a healthy and safe manner online.

10. THE MEANING OF THE WORD "FRIEND" IS CHANGING FOR OUR STUDENTS AND THIS CHANGE PUTS THEM AT RISK IN SEVERAL WAYS.

Ask an average teenager how many friends they have in their Facebook account and from some you may hear numbers between 200 and 500. "Friending" is a verb and for many of our students, some of their friends are complete strangers. We need to challenge them to think about what a friend is and consider the ways we typically value friends. Words like trust, love, support, and sharing come to mind. However, students' risks rise when they apply traditional real-life values to the "friendships" some of them develop online in sites such as Facebook.

We have Facebook accounts and actually see it as a wonderful and valuable resource. However, just because Facebook says that anyone 14 years or older CAN use Facebook, doesn't mean that they should. It isn't an age-appropriate or developmentally healthy place for our children and younger teens to hang out. Facebook is not working to protect our children and the laws in our country are terribly inadequate to safeguard our children online, in general. Not enough is being done to protect and educate children and teens against the risks that come from using the Internet and Facebook in particular. We (adults, parents, educators) need to do more.

In addition, during the last few years our schools have been welcoming an influx of a new generation of teachers. These younger teachers are typically more comfortable with technology because they have grown up with it. This presents some challenges as well. For instance, must independent schools consider setting policies for teachers regarding the use of social networks like Facebook? Should we set guidelines for the possible social interaction of our teachers with their students in sites such as Facebook? Many independent schools are currently debating these questions. Articles related to this topic make very plausible arguments for setting guidelines for teachers, as well as students.

—*South Florida Sun-Sentinel.* "Area teachers post questionable content on Facebook" by Stephanie Horvath (http://www.sun-sentinel.com/news/education/sfl-flpfacebook0601pnjun01,0,370501.story).

To read other articles such as this one, visit Google and enter the words teacher, Facebook, and content. ChildrenOnline.org produces a free monthly newsletter that is designed to keep educators and parents informed about the latest issues affecting children online and to address specific questions often raised by parents and teachers.

One final note: The Internet is constantly changing, as are the ways that kids are using it. From recent visits to some independent schools, we have learned of a rising interest about which we are very concerned. Some middle and high school students have begun to discover online live broadcast TV, known as "social broadcasting." BlogTV.com is one such site where a visitor is able to use a built-in video camera to broadcast him or herself live on the Internet. Anyone can stop by, enter a chat window, and anonymously interact with the person broadcasting. As you can imagine, without any controls, standards, or boundaries, this technology can have some serious negative consequences for some children and teens. For some of our students, using this technology can be irresistible, especially younger children who see themselves as being on real TV.

Doug Fodeman is co-director of ChildrenOnline.org and director of technology at Brookwood School, Manchester, MA 01944. He may be reached at *DFodeman@Brookwood.edu* or DougF@ChildrenOnline.org.

Marje Monroe, M.S.W, is co-director of ChildrenOnline.org. She may be reached at *MarjeM@ChildrenOnline.org.*

NOTES

ChildrenOnline.org offers innovative and comprehensive workshops on Internet safety and online education to students, parents, faculty, and administrators. Their approach combines a thorough understanding of Internet technologies, child development, and counseling, to focus on the impact of the Internet on the social, emotional, and language development of young people.

This article previously appeared on the web site of the National Association of Independent Schools at http://www.nais.org/resources/article.cfm?ItemNumber=151505.

Teacher-Librarian acknowledges the authors' permission to adapt and reprint this article.

Part V:

Leadership in the Learning Commons

The willingness of professionals and support staff in the school to consider change and take risks with the changing educational environment, technology, and unique student social networking behaviors is a key element in the establishment and sustainability of a Learning Commons concept. People make a difference. Leading people to that difference makes a difference.

In the lead article, Zmuda and Harada lay out the need for teacher librarians to develop a new role in order to make a fundamental impact on teaching and learning. What they say applies equally to other specialists in the school such as teacher technologists, reading coaches, counselors, or any other experts seeking more impact on what goes on in the classroom.

Throughout the articles in this section, a vision of collaboration and partnership between specialists and classroom teachers is explored and the idea that two heads are better than one is widely suggested. It is repeated that classroom teachers need not shoulder the entire burden of success; that partnerships and team efforts, so common in business and industry, are a natural in school environments. The authors in this section outline their efforts at change with the realization that their experiments are still developing.

FEATURE ARTICLE

librarians as learning specialists: moving from the margins to the mainstream of school leadership

IT IS OUR BELIEF THAT BUILDING AND DISTRICT LEADERSHIP *MUST* COME TO ENVISION THE LIBRARY AS INTEGRAL TO STUDENTS ACHIEVING THE MISSION OF THEIR RESPECTIVE SCHOOLS. WE DO NOT USE THE WORD *MUST* LIGHTLY: THE STAKES OF PREPARING STUDENTS FOR THE 21ST CENTURY WORLD HAVE NEVER BEEN GREATER.

One of the most common concerns teacher-librarians have shared with us across the country is the lack of understanding their administrators and their colleagues have about what is possible "if only" they were given the opportunity, the resources, and the support. Contrary to what some may believe, the lack of opportunity, resources, and support are not a personal attack nor is it a show of disrespect. The crux of the problem is that most administrators and staff fundamentally do not understand what is possible (despite many valiant efforts by teacher-librarians to explain it). They cannot separate out the librarian from the library because of minimal to no knowledge of the profession.

The reality of this information problem has been the focus of our writing as we address the library media specialist/teacher-librarian and administrator alike. After all, the work of the school is the work of the library. We urge the creation of a more focused job description and a more obvious set of collaborative partners in the architecture of schools. We build on research and literature that are renowned not only in the library profession but also seminal for administrators and teacher leaders.

In this article, we first present some of the critical problems facing schools and focus on the need to practice a mission-focused mindset that empowers school leadership teams to drive school improvement. In creating these teams, we propose that building administrators leverage the expertise of learning specialists, key among them, the teacher-librarian.

WHAT STATISTICS TELL US

The statistics on high school dropout rates presented at the National Education Summit on High Schools have been grim: "Today only 68 out of 100 entering ninth-grade will graduate from high school on schedule. Fewer than 20 will graduate on time from college. Meanwhile, 80% of the fastest-growing jobs will require some postsecondary education" (Education Trust, 2005, p. 3). According to the Bill and Melinda Gates Foundation, "every day nearly 3,000 of America's students drop out of high school. . . . Over the course of their lives, dropouts from a single year's graduating class cost the nation more than $325 billion in lost wages, taxes, and productivity" (n.d.).

Preparing students for the rigors of an information age requires not only getting them to earn a high school diploma but also enabling them to succeed in their further studies. Completion rates for students enrolled in postsecondary programs are equally troubling. David Conley (2007), who advocates for clarifying "standards for success" to prepare students for college-level tasks, reports: "The most recent data available show that only about 35% of students who entered four-year colleges seeking a bachelor's degree in 1998 had earned their degree four years later, and only 56% had graduated six years later" (p. 24). Conley largely attributes the low graduation rates to the complexity of the work required of them, the pace of the work, and the collaborative and communicative nature of the tasks. He quotes the National Research Council on college expectations:

> College instructors expect students to draw inferences, interpret results, analyze conflicting source documents, support arguments with evidence, solve complex problems that have no obvious answer, draw

by allison zmuda and violet h. harada

> "Preparing students for the rigors of an Information Age requires not only getting them to earn a high school diploma but also enabling them to succeed in their further studies."

> "A school that believes in rigorous and relevant student-focused learning also commits to a mission-centered mindset."

conclusions, offer explanations, conduct research, and generally think deeply about what they are being taught. (Conley, 2007, p. 24)

The expectations cited here contrast sharply with common practices in K–12 schools, such as teaching to standardized tests; covering curriculum topics at a breathtaking rate in order to meet all of the standards; and assessing students on the knowledge and skills they can recall based on a set of familiar problems, situations, and contexts. Mel Levine (2007) describes the cognitive toll of these practices on the development of the student:

> Many students emerge from high school as passive processors who simply sop up intellectual input without active response. Some passive learners, although able to scrape by academically, endure chronic boredom in school and later suffer career ennui. Their habit of cognitive inactivity can lead to mediocre performance in college and later on the job. (p. 19)

STUDENT-FOCUSED MISSION

To overcome the passivity of learners that Levine describes, 21st-century schools must embrace learning beliefs that produce engaged and sustained learning and develop skills of independence, problem solving, and teamwork. Students must constantly see the value of their work and feel a growing sense of efficacy. They must connect isolated facts and skills with big ideas and receive regular and user-friendly feedback to better understand goals and meet high standards. They must reflect, self-assess, and rethink ideas in a safe and supportive environment that fosters questioning assumptions (Wiggins & McTighe, 2007).

A school that believes in rigorous and relevant student-focused learning also commits to a mission-centered mindset. Mission both motivates and measures improved purpose because all stakeholders believe that the learner-based accomplishments they are in business to produce are challenging, possible, and worthy of the attempt. A mission-focused approach requires a constant analysis of whether daily practices are having the desired effects on student achievement. Such analysis will also uncover areas of misalignment where significant resources are expended to support the development of work that is tangential to established curricular goals.

The adoption and establishment of a set of learning principles is, therefore, critical to reform instructional practices that defy what we know to be true about how people learn. Every staff member must be held accountable (by their supervisors and by one another) to work in a way that will get the desired learning results. Every member practices such basic moves as:

- making the learning goals of the task/activity explicit to the students
- creating meaningful connections between the learning activity and the "real world" of the student and of professionals in the discipline
- providing regular, criterion-based feedback to students on the quality of their work and with regular opportunities to improve their work
- checking for understanding (and misunderstandings) early and often (Marzano, 2007; Schmoker, 2006; Wiggins & McTighe, 2007)

The significance of a powerful, consensus-driven mission statement and accompanying learning principles cannot be overstated. They provide the coherence, the alignment, the discipline, and the flow necessary for success. Staffs that practice mission-centered beliefs focus on a handful of improvement efforts, collaborate with one another to analyze student work as well as each other's instructional practice, and acquire new knowledge and skills, even if it means unlearning old ones.

TEAMING ON STUDENT-FOCUSED LEARNING

School-level leadership is essential in building a mission-centered culture. While current research confirms that effective building administrators are a necessary precondition to effective school reform programs, various studies also indicate that school leadership has shifted from a focus on single individuals to a team of individuals (Marzano, Waters, & McNulty, 2005).

The significance of a powerful, consensus-driven mission statement and accompanying learning principles cannot be overstated. They provide the coherence, the alignment, the discipline, and the flow necessary for success. Staffs that practice mission-centered beliefs focus on a handful of improvement efforts, collaborate with one another to analyze student work as well as each other's instructional practice, and acquire new knowledge and skills, even if it means unlearning old ones.

For formal leaders to nurture collaboration between learning specialists and staff, there must be strong internal accountability for student learning and a culture of trust. Doug Reeves (2006) outlines "essential truths" about this form of leadership:

- Employees in any organization are volunteers. We can compel their attendance and compliance, but only they can volunteer their hearts and minds.
- Leaders can make decisions with their authority, but they can implement those decisions only through collaboration.
- Leaders must leverage for improved organizational performance that happens through networks, not individuals. (p. 52).

Shared leadership is predicated on establishing and sustaining "purposeful communities with the collective efficacy and capability to develop and use assets to accomplish goals that matter to all community members through agreed-upon processes" (Marzano, Waters, & McNulty, 2005, p. 99). In such teams, members hold a shared belief that they can facilitate change. They leverage all available assets. They have well-articulated goals. They use processes that enable effective communication among members. We maintain that *learning specialists* are critical members of such school teams.

ROLE OF LEARNING SPECIALISTS

Who are learning specialists? They are partners with classroom teachers who play a central role in the continuous effort to improve the achievement of all students through the design, instruction, and evaluation of student learning. While learning specialists have worked in schools for years, these positions have multiplied with the advent of rigorous content standards and related state assessments as well as research on effective staff development.

Learning specialists are often entrusted with coordinating a program; designing enhanced services in a curricular area; or providing specialized services to students, teachers, and even parents. Most learning specialists have a teaching license as well as additional certification or credentials in a specialized area. Frequently, they are referred to as *informal leaders* or *instructional leaders* in a *distributed model* because they typically are not required to have the administration certification needed to supervise teaching personnel. A characteristic that distinguishes the learning specialist from the classroom teacher is that the person frequently has no official full assignment of students. Learning specialists in schools can include (but are not limited to) the following: reading, writing, mathematics, fine arts, and science coaches; instructional technology coordinators; and teacher-librarians. Teacher-librarians have been staff members in K–12 schools since the early 1900s (Morris, 2004), while other positions, including reading specialist positions, have been added in many schools since the 1960s (Bean, Swan, & Knaub, 2003).

While learning specialists may have highly specialized roles, the common characteristics that shape their jobs make them natural partners in the work of formal leadership to raise student achievement. Working in separate silos merely diminishes each learning specialist's efforts and weakens every individual's ability to effect improvements in teaching and learning. In short, isolationism further devalues the specific service rendered. It is good business, therefore, to work strategically as a cooperative unit of professionals targeting goals that might be met by leveraging the resources and talents of the team.

Collegiality denotes the ability of staff members to work with one another in the analysis of curriculum documents, assessment results, and instructional strategies without getting mired in personal politics. This key school-level factor requires a *constructive process* where staff members cooperatively determine how to replicate those teaching practices that result in the desired student learning. Doug Reeves (2004) states that the "difference between malaise and effectiveness is the collective will of the faculty to focus on their strengths, to ask one another questions, and to take responsibility for their professional growth and the achievement of their students" (p. 38). Only through this type of constructive process does craft knowledge truly begin to flourish.

Deanna Burney (2004) further defines craft knowledge as "research knowledge that is informed by practice, that is codified, tested, and shared" (p. 527). She elaborates:

People learn by watching one another, seeing various ways of solving a single problem, sharing their different "takes" on a concept or struggle, and developing a common language with which to talk about their goals, their work, and their ways of monitoring their progress or diagnosing their difficulties. When teachers publicly display what they are thinking, they learn from one another, but they also learn through articulating their ideas, justifying their views, and making valid arguments. (p. 528)

The goal is *not* to increase collaboration; the goal is to improve student performance. The goal is *not* to force staff to attend professional development; the goal is for them to improve their practice in order to improve student performance. The goal is *not* to garner more respect for the learning specialists; the goal is for the interactions between learning specialists and staff to help the system improve its overall performance.

Consensual change occurs when staff distinguishes between what they *like* or *prefer to do* from *what actually works*. When school teams collaborate to clarify the relationship between the design and the effect on achievement, they witness positive and constructive change at staff meetings, in classrooms, and in individual staff development sessions. Deborah Meier maintains:

> The kinds of changes required by today's (education reform) agenda can only be the work of thoughtful teachers. Either we acknowledge and create conditions based on this fact, conditions for teachers to work collectively and collaboratively and openly, or we create conditions that encourage resistance, secrecy, and sabotage. (quoted in Wagner, 2003, p. 101)

Learning specialists have the unique position to affect classroom-level practice in significant ways because of their student-centered mindset and content and pedagogical expertise. As members of the leadership team, they can create the conditions for internal accountability so that staff members hold one another accountable for student achievement, staff development, and coherence of leadership efforts.

TEACHER-LIBRARIANS AS LEARNING SPECIALISTS

Teacher-librarians are strategically positioned to be influential members of school leadership teams. Turner and Riedling (2003) contend the "greatest cause for optimism is the fact that library media specialists are in the right place at the right time to play a significant role in the transformation of teaching that must occur as K–12 education is impacted by the revolution in telecommunications and information technologies" (p. 232). As learning specialists, they can grow the expertise of the teaching staff through the collaborative tasks they complete together, from the staff development workshops they design, and from the modeling they do in the library-classroom.

In our extensive conversations and observations of the work being done by teacher-librarians across the nation, we were inspired by evidences of best leadership practices. Clearly, there are impressive examples of teacher-librarians who believe that student learning is the winning priority of their programs. We noted common threads in the actions of these teacher-librarians that confirm Charlotte Danielson's (2007, pp. 124–131) observations:

- They volunteer for leadership roles within the school and district to articulate the needs of students in information fluency within the school's academic program.
- They articulate and communicate student-focused goals for the library program that are highly appropriate to the situation in the school and to the age of the students.
- They are knowledgeable of resources available for students and teachers and actively seek out new resources to enrich the school's program.
- They initiate collaboration with teachers in the design of instructional lessons and units that result in coteaching.
- They are always searching for innovative ways to use current and emerging technologies to enhance the learning experience for students and teachers.
- They create learning environments in which students engage in inquiries that challenge them to think critically and act creatively and responsibly.
- They interact with students and teachers in ways that are highly respectful, reflecting genuine caring and sensitivity to students' cultures and levels of development.

Together with the other professionals in the school, teacher-librarians practice research-based pedagogy by

- monitoring student learning and making adjustments in "real time" without compromising students' opportunity to learn
- designing instruction in "small chunks" without compromising students' ability to see the "big picture" or to become overly dependent on the teacher to make meaning for them
- personalizing instruction to fit the needs of each learner without compromising belief that all students can achieve high expectations
- incorporating student interests into curriculum, assessment, and instructional design without diluting the rigor or focus on learning goals
- developing a "team mindset" among learners without compromising the ability to truly get to know each person individually
- inspiring students to find schoolwork meaningful and challenging without sending mixed messages through the assignment of low-level worksheets and recall activities
- challenging learners to pursue inquiries with no clear answer and problems that they have never encountered before without rushing through the experience.

In schools where collaborative professional communities flourish, teacher-librarians are respected teaching partners who positively affect student learning based on observable indicators such as:

- rebuilding assessment tasks or instructional experiences that enhance rigorous learning;
- evaluating student work to determine the extent to which their collaboration improved achievement and how that informs future collaborations;
- exchanging feedback and guidance with teachers on ways to strengthen practices that raise student achievement in information literacy and technology;
- receiving the principal's full endorsement to participate in key committees, budget decisions, and staff development opportunities.

The bottom line is this: teacher-librarians view their work as "the school's work," not just because the physical space and resources are shared by all, but because the significance of the learning that is conducted in the library is at the heart of the school's purpose. This mission-centered mindset—preparing all students to be successful in a 21st-century world—gives teacher-librarians the authority to work as partners in the design and evaluation of student learning. The future viability of the library depends upon the willingness of teacher-librarians to hold themselves, their students, and their colleagues accountable for creating a learning environment and learning experiences that accomplish the curricular goals delineated in AASL's *Standards for the 21st Century Learner* (2007).

Change of this magnitude requires not only rethinking the library media center's mission but also reassessing current practice and reinventing what teacher-librarians accomplish as learning specialists. In his foreword to *Librarians as Learning Specialists* (Zmuda & Harada, 2008), Grant Wiggins says that for him, the library "has always been such a revealing barometer" of a purposeful and healthy school. He states that the library is a "window into how well the entire staff understands learning and honors best practice." For libraries to be these windows, they must be more than physical warehouses for resources. They must be "inquiry laboratories" where students and instructional teams explore problems, seek answers to questions, and pursue personal needs for information (Kuhlthau, Maniotes, & Caspari, 2007, p. 63).

LEARNING SPECIALISTS IN ACTION

Here are brief snapshots of teacher-librarians striving to achieve a student-focused mission.

CREATING INQUIRY ENVIRONMENTS

Anna, a teacher-librarian in an urban elementary school, has transformed her library into what she calls an "exploratorium." When you enter the facility, the first thing you see is a "wonder tower," a cardboard pyramid that is covered with questions generated by students. Youngsters are encouraged to contribute questions they are curious about. In turn, other students are invited to post responses and cite their information sources. The exploratorium has low bookshelves that are usually covered with intriguing realia and artifacts—students are challenged to figure out what they are and how they might be used. Last month, for example, Anna displayed a variety of kitchen utensils from colonial times that she borrowed from a lending collection of a nearby museum. She held a contest for students to guess their various uses. Anna has also worked with the school's curriculum coordinator to plan simple mini-inquiry centers in her exploratorium. Each center (on a small table) focuses on a key curriculum-related question and includes a range of resources that help students to explore the question.

USING TECHNOLOGY TO TRANSFORM LEARNING

Ryan, a teacher-librarian in a rural middle school, has created a learning hub that uses a range of tech tools to motivate adolescents. He and the school's reading coach have established a cyber-book club where members conduct electronic discussions. With input from faculty and students, Ryan is building a library web site that includes a range of subject-specific search tools, e-books, and online databases. He has created a blog to highlight library news and upcoming events. With the support of his faculty and the school's technology resource teacher, he has taught students to do Podcasting as well as how to work with wikis. What is important: He has also held informal information and training sessions on these various resources and tools for teachers, administrators, and parents.

ASSUMING LEADERSHIP IN SCHOOL TEAMS

Mary and Sam are teacher-librarians at a suburban high school where students and teachers are organized in a range of learning academies. The curriculum is interdisciplinary, and students work in teams to conduct research and to design and develop projects based on their findings. Along with supporting the students and faculty with resources, both teacher-librarians have volunteered to assist with professional development activities. In this capacity, they have taken the lead training teachers to develop essential questions and design assessment tools for benchmark tasks in the research process. Teaming with the school's technology resource coordinator, Mary and Sam have started to explore the use of Second Life as a virtual gallery for student-produced artifacts. They are also working with the teachers on curriculum maps that reflect the integration of the *Standards for the 21st Century Learner* (AASL, 2007) with content standards.

CONCLUSION

For teacher-librarians and other potential change agents to move from the margins to the mainstream of their schools, Marzano and colleagues (2005) contend they must wrestle with and act on hard questions: Do I systematically consider new and better ways of teaching? Am I willing to lead change initiatives with uncertain outcomes? Do I consistently try to operate on the cutting edge versus the center of the school's competence? Importantly, if administrators wish to empower teacher-librarians and other learning specialists in their schools to assume a mantle of shared leadership, they must legitimize their role in designing and implementing curriculum, instruction, and assessment activities.

Shared leadership holds the bright promise of building and sustaining a professional culture of best practice. In schools where this concept of purposeful community is alive, we find students rigorously engaged in the construction of knowledge and the communication of thinking.

REFERENCES

American Association of School Librarians (AASL). (2007). *Standards for the 21st century learner.* Retrieved May 2, 2008, from www.ala.org/ala/aasl/aaslproftools/learningstandards/standards.cfm.

Bean, R. M., Swan, A. L., & Knaub, R. (2003). Reading specialists in schools with exemplary reading programs: Functional, versatile, and prepared principals and reading specialists in schools with exemplary reading programs were asked about the perceived role of the reading specialist. Results indicate that training programs for specialists should include more leadership skills. *The Reading Teacher, 56*(5), 446+. Retrieved May 2, 2008, from www.questia.com/PM.qst?a=o&td=5000631482.

Bill and Melinda Gates Foundation. (n.d.) "Education Fact Sheet." Retrieved May 2, 2008, from www.gatesfoundation.org/UnitedStates/Education/RelatedInfo/EducationFactSheet-021201.htm.

> "The bottom line is this: teacher-librarians view their work as 'the school's work,' not just because the physical space and resources are shared by all, but because the significance of the learning that is conducted in the library is at the heart of the school's purpose."

> "Shared leadership holds the bright promise of building and sustaining a professional culture of best practice."

Burney, D. (2004). Craft knowledge: The road to transforming schools. *Phi Delta Kappan, 85*(7), 526-531. Retrieved May 2, 2008, from www.questia.com/googleScholar.qst?docId=5002090878.

Conley, D. T. (2007). The challenge of college readiness. *Education Leadership, 64*(7), 23-29.

Danielson, C. (2007). *Enhancing professional practice: A framework for teaching* (2nd ed.). Alexandria, VA: Association for Supervision and Curriculum Development.

The Education Trust. (2005). *Gaining traction, gaining ground: How some high schools accelerate learning for struggling students.* Retrieved May 2, 2008, from www.ecs.org/html/Document.asp?chouseid=6661.

Kuhlthau, C. C., Maniotes, L. K., & Caspari, A. K. (2007). *Guided inquiry: Learning in the 21st century.* Westport, CT: Libraries Unlimited.

Levine, M. (2007). The essential cognitive backpack. *Education Leadership, 64*(7), 16-22.

Marzano, R. (2007). *The art and science of teaching: A comprehensive framework for effective instruction.* Alexandria, VA: Association for Supervision and Curriculum Development.

Marzano, R. J., Waters, T., & McNulty, B. A. (2005). *School leadership that works.* Alexandria, VA: Association for Supervision and Curriculum Development.

Morris, B. (2004). *Administering the school library media center* (4th ed.). Westport, CT: Libraries Unlimited.

Reeves, D. (2004). *Accountability for learning: How teachers and school leaders can take charge.* Alexandria, VA: Association for Supervision and Curriculum Development.

Reeves, D. (2006). *The learning leader: How to focus school improvement for better results.* Alexandria, VA: Association for Supervision and Curriculum Development.

Schmoker, M. (2006). *Results now: How we can achieve unprecedented improvements in teaching and learning.* Alexandria, VA: Association for Supervision and Curriculum Development.

Turner, P. M., & Riedling, A. M. (2003). *Helping teachers teach: A school library media specialist's role* (3rd ed.). Westport, CT: Libraries Unlimited.

Wagner, T. (2003). *Making the grade: Reinventing America's schools.* New York: RoutledgeFalmer.

Wiggins, G., & McTighe, J. (2007). *Schooling by design.* Alexandria, VA: Association for Supervision and Curriculum Development.

Zmuda, A. & Harada, V.H. (2008). *Librarians as learning specialists: Meeting the learning imperative for the 21st century.* Westport, CT: Libraries Unlimited.

Allison G. Zmuda is an education consultant who has worked with schools throughout the United States and Canada. You may contact her at *zmuda@competentclassroom.com*.

Violet H. Harada is a professor in the University of Hawaii's Library and Information Science Program. You may contact her at *vharada@hawaii.edu*.

Feature articles in *TL* are blind-refereed by members of the advisory board. This article was submitted June 2008 and accepted September 2008.

FEATURE ARTICLE

SCHOOL LIBRARY 2.0: FROM THE FIELD

information and technology literacy

EDITOR'S NOTE: HOW DOES ONE PERSON REVOLUTIONIZE AN ENTIRE DISTRICT BOTH IN TECHNOLOGY AND IMPROVED INSTRUCTIONAL DESIGN? WHEN BILL DERRY ACCEPTED THE POSITION AS COORDINATOR OF INFORMATION AND TECHNOLOGY LITERACY IN WESTPORT, CT, HE DID SO WITH THE ENCOURAGEMENT TO BRING MAJOR CHANGE TO HIS MULTI-SCHOOL DISTRICT. HERE IS HIS PROGRESS REPORT THAT IS WORTH READING BECAUSE OF THE ORGANIZATIONAL STRUCTURE HE SET UP TO MOVE US ALONG IN THE JOURNEY TOWARD EXCELLENCE.

Our mission for the Westport, CT, public schools is "to help students acquire the attributes necessary to be successful in the complex, technological, information-based, and rapidly changing 21st-century world." Since 2001, this mission has guided the development of our Information and Technology Literacy (ITL) curriculum, as well as the creation of implementation strategies to ensure that all staff members take ownership of the ITL curriculum. Our mission makes it logical and necessary for all members of our community to work together to utilize ITL skills and strategies in our daily teaching.

In this article, I chronicle some of our group's successes and challenges. As coordinator of ITL for 2 years and a former elementary teacher-librarian in the district for 6 years, I am writing about initiatives made possible by a large group of people, including the superintendent, assistant superintendent for curriculum and instruction, the present and previous directors of technology and their staffs, all school administrators, all library media and technology staff, all teachers on the curriculum writing committees, and each school's ITL committee members.

ITL COMMITTEES

The first implementation strategy involved the creation of ITL committees at each school. These committees are made up of a cross-section of grade levels and roles found at each school and chaired by teacher-librarians, technology teachers, classroom teachers or administrators with the goal of moving ownership of the ITL curriculum to everyone (http://tinyurl.com/6qal2h). We have approximately 580 professional district staff (including teachers, administrators, and specialists of all kinds) in our 8 schools, and currently on our 8 ITL committees, there are 92 members. This means that 16% of the entire professional staff is having a regular conversation and taking action on integrating ITL skills and strategies in meetings that occur at least 4 to 12 times each year. Each ITL committee determines the appropriate goals and objectives for its school and maintains a wiki to share agendas, notes and ideas between schools (http://tinyurl.com/63z9ug). Representatives from each ITL committee meet as an ITL steering committee three times each year to share successes and challenges and take part in professional development. This year, we had Dr. Don Leu

> We have approximately 580 professional district staff (including teachers, administrators, and specialists of all kinds) in our 8 schools, and currently on our 8 ITL committees, there are 92 members. This means that 16% of the entire professional staff is having a regular conversation and taking action on integrating ITL skills and strategies in meetings that occur at least 4 to 12 times each year.

bill derry

22 | TEACHER LIBRARIAN 36:1

from the University of Connecticut present his research on "New Literacies."

PROFESSIONAL DEVELOPMENT

Another approach has been the creation of a 4-day ITL Summer Institute designed to educate staff on the use and integration of ITL skills and strategies to create engaging learning activities that challenge students to think at a higher level. Participants register months in advance and are paid to attend. The 2007 keynote and workshop coach for 3 days was David Loertscher. He presented strategies to 42 participants on "high think" learning activities from the book he coauthored with Carol Koechlin and Sandi Zwaan, *Beyond Bird Units* (2007). Participants also took part in activities that utilized a common planning tool, wikis, social networking (Ning), digital storytelling (PhotoStory and Podcasting), interactive whiteboards (Smartboard), and subscription databases. Every school continued in-house professional development throughout the school year, increasing the impact of the Summer Institute many times over. The institute was so successful that the 2008 institute had 72 participants with 8 people on the wait list.

This year, David Warlick provided the keynote on redefining literacy in the 21st century and supported the development of blogging. Workshops were provided on most of the topics from the 2007 institute, with sessions on "new literacies" and a live presentation from Kathy Schrock in Second Life. All participants experienced the constructivist world of a MUVE and traveled into Second Life with their own avatars. We anticipate that this year's ITL committees will use the institute as a springboard for determining goals, objectives, and school-based professional development throughout the 2008-2009 school year.

The institute was part of a larger district conversation on identifying specific skills that are identifiable and necessary in the 21st century. For the institute's "big think," each school presented a 1-minute defense of the most critical skill for students to master in the 21st century, utilizing technologies that best presented the message.

The two challenges of this initiative are to be able to differentiate instruction for the wide range of abilities attending and to grow the budget to fit the increasing numbers interested in attending (http://tinyurl.com/6xrv3f and http://tinyurl.com/692lue).

ASSESSMENT

Another strategy for ITL implementation involved adding Research and Technology to the fifth-grade report cards in the 2007-2008 school year. The grades were given twice by each classroom teacher with the idea that there would be collaboration with the teacher-librarian and technology teacher at each school. This action was a response to the question "How can we determine what our elementary students leaving fifth grade know and are able to do in terms of their ITL competencies?" A year-long conversation with principals, fifth-grade teachers, teacher-librarians, and technology teachers led to the reduction of 259 ITL performance objectives to 18 research and 14 technology "expected outcomes." Two rubrics were created by teacher-librarians and technology teachers and reviewed by classroom teachers and administrators to help with the assessment of student research and technology work. (All these documents are available at http://tinyurl.com/5jd2ya; use key "westportitl.")

There are at least two big challenges in this initiative. We have fought to integrate ITL into subject areas, so what does it mean when you isolate research and technology as separate subject areas on a report card? Also how do we ensure that teacher-librarians and technology teachers do, in fact, play a role in cooperatively planning, implementing, and assessing ITL competencies? As you would expect, each school and each team and each individual interprets and executes this process at various levels. This year we will focus on the process and attempt to share successes and challenges. Ultimately we hope to influence the refining of the ITL curriculum in the middle schools and high school.

DIGITAL VIDEO COMMUNICATIONS

Another strategy that has evolved from the board approval of our ITL curriculum in 2004 involves the implementation of television studios in all schools. The high school and two middle schools had their own functioning television studios, and in 2006, one elementary school came onboard. In the 2007-2008 school year, three new television studios were added, sending their digital communications over the Internet rather than coaxial cable. Next year, the last school will have a television studio. In all cases, a teacher-librarian and/or technology teacher (often working with one or more classroom teachers) is critical to its success. This initiative provides more challenges than can be detailed in this article, but three include: revising the ITL curriculum to provide sequenced media literacy objectives, managing the numbers of students wanting to produce television programming, and creating the environment to accept and utilize the evolution from the broadcast paradigm of the 20th century to the multicast paradigm of the 21st century.

NEW SYNCHRONOUS AND ASYNCHRONOUS TOOLS

Our last strategy involves keeping abreast of new technologies and utilizing them to improve teaching and learning. Last year, some of our ITL committees provided professional development opportunities and utilized many new synchronous tools, including instant messaging, live webinars, and Skype, to bring lectures to people from

---

We anticipate that this year's ITL committees will use the institute as a springboard for determining goals, objectives, and school-based professional development throughout the 2008-2009 school year.

other countries and towns to instantly share information between participants. Several asynchronous tools, such as wikis, blogs, and e-mail, were frequently used. This year, our district is implementing Blackboard, which promises to revolutionize how we manage and deliver course content and new systems of communication between students and teachers. It is clear that each school's ITL committee will focus heavily on professional development in this area.

CONCLUSION

With our ITL committees, ITL Summer Institutes, fifth-grade assessments, television studios in all schools, and rapid increase in the K-12 use of Web 2.0 tools, it is clear we have made progress in implementing our ITL curriculum. These changes have challenged us to review our acceptable use policies and our learning activities related to Internet safety and the responsible and ethical use of information. We are aware that our ITL curriculum of 2004 is ready for revision, and we must prepare our committees for this task. Those who are using and flexibly adapting some of the Web 2.0 tools to construct new paradigms for teaching and learning are poised to provide leadership in each of these areas. A new kind of collaboration (related to social networking, interactive Web-based tools, and decentralized authority based on group consensus) has emerged. Those teachers who connect to this new collaborative model seem best able to teach with and model these skills within their subject areas. They are the change agents who are ready to rapidly synthesize, adapt, and apply new ITL resources to make learning more engaging and authentic.

In an attempt to ensure that all teacher-librarians are equipped to better manage change, either through collaborating with those who are change agents or taking the role of "change agent," some of our professional development this year will focus on the identification and development of those skills. A wiki has been set up to support this initiative (http://managingchange.pbwiki.com). Please request to join this wiki if you would like to participate in the conversation.

REFERENCES

Loertscher, D., Koechlin, C., & Zwaan, S. (2007). *Beyond bird units: Thinking and understanding in information-rich and technology-rich environments.* Salt Lake City, UT: Hi Willow Research.

Bill Derry is coordinator for Information and Technology Literacy for the Westport, Connecticut public schools. He can be reached at *bill_derry@westport.k12.ct.us*.

"Our mission makes it logical and necessary for all members of our community to work together to utilize ITL (Information and Technology Literacy) skills and strategies in our daily teaching."

"These changes have challenged us to review our acceptable use policies and our learning activities related to Internet safety and the responsible and ethical use of information.....A new kind of collaboration has emerged.....In an attempt to ensure that all teacher-librarians are equipped to better manage change,...,some of our professional development this year will focus on the identification and development of those skills."

FROM THE BRAIN TRUST

Technology Leadership: Kelly Czarnecki

TL's Editors

While not a teacher-librarian, Kelly Czarnecki works often and collaboratively with children and youth that attend schools in the Charlotte, NC area.

Kelly is a Technology Education Librarian at ImaginOn, a youth playground of sorts, in Charlotte (www.imaginon.org). She was given the Mover and Shaker award by *Library Journal* in 2007 for her work with youth and technology. She writes the Gaming Life column for *School Library Journal* and is working on a book on gaming in libraries to be published in 2010; another book is forthcoming on Mashups in the library. Her latest book *Digital Storytelling in Practice* was published in October. Here are her thoughts:

TL: How does your work relate to school librarianship?

Czarnecki: I have always worked with teacher-librarians, even when I was a librarian in a public library. I think it is important to collaborate and share our ideas, resources, and expertise since we are often serving the same population. When I first started partnering, it was to give booktalks on a monthly basis at the local middle and high schools in central Illinois. Coming to Charlotte and ImaginOn, the partnerships mostly involve our technology resources.

At ImaginOn, we have ReadyANIMATOR, www.readyanimator.org, that allows users to create movies and stop motion animation. It was built by John Lemmon from John Lemmon Film Studios in Charlotte and uses an iMac with such software as iCanAnimate, www.kudlian.net/products/icananimate/, GarageBand, http://www.apple.com/ilife/garageband/, iMovie, and iDVD. It is definitely an important tool for digital storytelling and portable filmmaking if your budget allows.

I have also partnered with school libraries in promoting events such as Teen Read Week, Teen Tech Week, and summer reading. I am currently working with the Northwest School of the Arts in an after school board, card, and video gaming/tutoring program.

TL: How have you and the Charlotte & Mecklenburg County libraries become a part of the evolution of teaching and learning?

Czarnecki: Over time, my library system has become part of the teaching and learning process by partnering with schools on various projects. We have taken the ReadyANIMATOR to schools and teachers use it to assist in their classes on filmmaking. This year we are a partner in an Arts & Science Council grant with a local charter school. Students K-5 will learn how to make slideshows and videos using the photographs they take and video footage they record. We will have the opportunity to assess how students and teachers think of the library differently (if they do), after the project.

TL: What can the teacher-librarian draw from what you (and the library) have done to help connect to the teaching and learning process?

Czarnecki: Become the leaders in showing what can be done with technology in the classroom. Sometimes we might not have time to learn things on our own but if we have leaders showing us the way, we might be more likely to delve in and see that it is not so scary.

I think today's teacher-librarian and teachers are linking in fantastic ways, particularly with technology skills and support. Joyce Valenza in Pennsylvania (www.sdst.org/shs/library/jvweb.html) frequently writes for *School Library Journal* and is doing amazing work at her school. Peg Sheehy in New York is an avid user of Teen Second Life to digitally engage students (http://ramapoislands.edublogs.org/).

We also need to consistently tell our stories and make public the effects we are having on our youth. It becomes a powerful statement that we are still relevant and central to the learning and development of our community.

I think it is essential for teacher-librarians to get their story out and get help to shine a spotlight on their work so others will hear about their work and replicate it. Starting a blog, writing an article for *Teacher Librarian* or other journals in the school library field are also some ways to start exposing more of what can be done by teacher-librarians.

TL: How has technology shaped what you do on a day-to-day basis at work to serve library users? And, from your perspective, how has technology influenced library work today?

Czarnecki: The 21st Century skills movement has influence on what is happening in school libraries today. In order to survive, young people need to know how to use technology for collaboration, play, critical thinking, negotiation, and more. Teacher-librarians are in a great position to be at the forefront of that movement and to help show youth how to use the tools effectively.

Each issue of *Teacher Librarian* will feature someone with expertise in the field of school librarian work and education. Featured will be professionals that have made a difference in the field.

The questions asked of each professional for this column will always include:

Reflection on their work in school librarianship, their ideas about technology in librarianship, and how teacher-librarians can become central to the teaching and learning process.

Do you have someone you would like to nominate for this column? Do you have a question you would like us to ask? Send that information with a brief paragraph about this professional to *b.marcoux@verizon.net* (Betty Marcoux).

Technology is part of everything I do at the library in providing services to our youth. However, we still need to find other creative ways to engage their attention and for that reason we have changed the way we offer technology workshops from when we first opened. We are doing a lot less 'lecturing' on our part to giving hands-on and interactive instruction. We always learn a lot from the youth and strive to put them in positions of leadership where they often help guide their peers. And sometimes that just naturally happens that they become instructor.

We also work to make technology more a part of our library building and system, which inspires the activities and services we offer. We have created such things as 'take out kits' so other branches can reserve a program on robotics for example, with all the equipment they need to get started.

TL: Is there something you see in the future of library and school library work that says that the ImaginOn is a concept for our future?

Czarnecki: Partnerships are definitely the key, especially at ImaginOn, which shares a building (and much more!) with the Children's Theatre of Charlotte. We are constantly working together. We develop technology and other programs related to the theatre performances for example.

I think the teaching profession is very similar in the sense that community partners can definitely make a positive effect on knowledge. With so many options for technologies available to 'bring' people inside the classroom, developing partnerships doesn't have to be a financial hardship.

TL: If you were to enter library school or public librarianship today, what would you find most interesting and why?

Czarnecki: I find that much reading has evolved from print format to so many other formats and ways of telling our stories. I think it's most interesting because people learn in different ways and may prefer one format at one time and another format for another time. It's interesting to see how youth move in and out of formats depending on their needs. Exposing ourselves to different formats helps us to remain relevant and able to understand why one way of reading or learning doesn't work for all and doesn't need to.

I think the opportunities to become an influential member of the professional community through teaching and writing are great for librarians as well. It's such an open community to share and build upon each other's knowledge. It's definitely a growing organism!

TL: What are some constants you see that remain from when you first began your work to today's profession? How does this relate to school librarianship?

Czarnecki: I believe the library is still that third place between home and school for many youth and that has not changed. Being a welcoming, up-to-date environment increases the chances of many youth using this place.

TL: School and public libraries are working in closer partnerships more than ever before. What is the best way for a librarian (public and school) to become equipped for today's library work?

Czarnecki: I recommend that you network online with professional communities, and that you network online within the school and with the students as much as possible. We can all learn so much from each other and using the many Web 2.0 tools available helps us to get at that information in a way that we never could before.

I think the teaching profession is very similar in the sense that community partners can definitely make a positive effect on knowledge.

FEATURE ARTICLE

SCHOOL LIBRARY 2.0: FROM THE FIELD

three heads are better than one: the reading coach, the classroom teacher, and the teacher-librarian

EDITOR'S NOTE: THE MAJOR DISCOVERY IN THIS ACCOUNT IS THAT THE SPECIALIST AND TEACHER-LIBRARIAN COLLABORATE WITH THE CLASSROOM TEACHER ON A LEARNING UNIT AND THEY DO SO TOGETHER. THIS IS NOT JUST PASSIVE OR FROM SOMEWHERE IN THE BACKGROUND, BUT ACTIVE COLLABORATION. THEIR WORK WITH ESL LEARNERS IS REMARKABLE AND DEMONSTRATES WHAT HAPPENS WHEN ESL YOUNGSTERS SUDDENLY FIND MEANING IN WHAT IS BEING ASKED OF THEM.

Winnie Porter is the teacher-librarian, Christopher Lamb is the Reading First coach, and Carol Lopez is the teacher in this collaboration exercise. Carol has been the 3rd grade Spanish bilingual teacher for 20 years and was excited to introduce the use of technology in her class. Christopher has worked with Carol for 3 years in his capacity as a coach; they have a good working relationship. This was Winnie's first year working as a teacher-librarian. Rowena Tong, the technology teacher, was part of the team and provided technical assistance.

What follows is a report of each participant's perception of the process.

Christopher: This experience was very exciting for us. I was able to collaborate with both a teacher and a teacher-librarian. In my current position as a reading first coach, I work closely with classroom teachers, assisting them in maximizing the effectiveness of their language arts instruction. Thus, this project represented a melding of my current duties and schooling.

As members of the school technology team, Winnie and I have been attending a series of technology workshops this year. We were given the mandate of incorporating the technology presented to us into actual lessons or units of study at our school sites. The third-grade students were preparing to learn how to write summaries of stories they had read. We decided to join forces to enrich this process.

Winnie: We took a traditionally boring assignment that students do periodically without enthusiasm. Introducing the technology turned it into an exciting, fun learning experience.

Christopher: We would focus on summary writing, an area I already assist teachers with as part of my regular duties. We worked together to utilize our technology training to help the students create an audiovisual summary.

Winnie: All three of us have worked as immersion/bilingual classroom teachers in the past, so we are fluent in Spanish. This is an important element to this project as all the kids in Carol's class are English-language learners.

Carol: This was the first year the children were developing their skills writing in English. Because the project took place toward the end of the year, they were able to produce more stories and choose from their favorite. They were also thrilled to be able to see themselves on the computer reading their final published pieces.

Christopher: This work took place in the classroom and school library. All three of the adults involved worked with the students together. The students understood that one of the goals of the project was to record a fluent reading of their summary and practiced over and over again, which is quite atypical. They were very engaged. There wasn't a single behavioral issue that surfaced in any of the sessions we cotaught. I am sure some of the excitement was due to the novelty of the new tools, but this only highlights the importance of intro-

christopher lamb, winnie porter, and carol lopez

SCHOOL LIBRARY 2.0: FROM THE FIELD

ducing new strategies and tools to promote student engagement. The excitement of the students rubbed off on the teachers!

Winnie: Working with a program in which the kids could record and hear their voice was extremely valuable. English-language learners need to hear their voice. The students realized on their own from the beginning that they needed to improve their oral fluency. After hearing themselves a couple of times, they self-corrected, slowed down, paid attention to punctuation, and improved their articulation. It has always been my experience that students hate to read their writing out loud and have to be forced to do so. In this project, they willingly read their pieces over and over without any teacher coercion.

Carol: The children always read their finished writing to the class through the author's chair, but this format took that one step further. They worked on fluency throughout the year by reading to partners or parents, so being able to see and hear themselves read aloud really showed them what they needed to work on without any teacher having to tell them.

Winnie: For all of us, the project brought back the joy of teaching.

Christopher: Through working together, we were able to ensure that student learning went well beyond the relatively simple task of writing a summary. Students were able to focus on fluency, a factor for all but of particular importance to English-language learners. They were able to utilize newly acquired technical skills. They were able to work together to critique each others' presentations.

This project illustrated for us the potential power of collaboration among the various staff members of a school. Each of us was able to contribute to the project. Because of my work as a literacy coach, I am very familiar with the demands of summary writing. I was able to work with the teacher and class to help them understand the key features and purposes of a summary. Ms. Lopez was familiar with the particular needs and strengths of the students in her class. Winnie is much more familiar with Macs than I am and set up the files for the students to save their work. The three of us, who are Spanish speakers, were then able to offer assistance to the students while they worked.

Carol: This class was such a hard working class and being chosen for this project made us all very proud. They were excited to use the technology, and it made my job easier because they were motivated to produce good writing.

Christopher: I feel very energized by this project and look forward to continuing this type of collaboration next year. We plan to do a presentation to the principal and staff regarding our project, as a means of educating them about the potential role of the library as the "hub" of the school.

Winnie Porter is a teacher-librarian at Paul Revere Elementary School, San Francisco, CA. She may be reached at *peruwinnie@yahoo.com*.

Christopher Lamb was the reading first coach at Paul Revere. He is now a teacher-librarian at two public elementary schools in San Francisco: Alvarado and Dr. Charles Drew. He may be reached at *clambo1212@yahoo.com*.

Carol Lopez is the teacher of the third-grade Spanish-bilingual class at Paul Revere and in September became the Spanish Immersion teacher. She may be reached at 415.695.5974.

"The students understood that one of the goals of the project was to record a fluent reading of their summary and practiced over and over again, which is quite atypical."

"Through working together, we were able to ensure that student learning went well beyond the relatively simple task of writing a summary. Students were able to focus on fluency, a factor for all but of particular importance to English-language learners. They were able to utilize newly acquired technical skills. They were able to work together to critique each others' presentations."

I feel very energized by this project and look forward to continuing this type of collaboration next year. We plan to do a presentation to the principal and staff regarding our project, as a means of educating them about the potential role of the library as the "hub" of the school.

FEATURE ARTICLE

SCHOOL LIBRARY 2.0: FROM THE FIELD

advanced contemporary literacy: an integrated approach to reading

EDITOR'S NOTE: IT IS NOT OFTEN THAT THE LIBRARY IS WRITTEN DIRECTLY INTO THE CURRICULUM, BUT SHARON SWARNER WAS ON THE RIGHT COMMITTEE AT THE RIGHT TIME. SHE WAS ABLE TO CONVINCE TEACHERS THAT REPLACING TIRED NOVEL STUDIES WITH HIGH-THINK AND ACTIVE-LEARNING EXPERIENCES WOULD BRING NEW LIFE INTO A GROUP OF BORED MIDDLE-SCHOOLERS' LIVES. THE AWAKENING TO HIGHLY COLLABORATIVE CLASSROOM/LIBRARY LEARNING EXPERIENCES WAS A MAJOR STEP FORWARD AND, VARIOUS SPECIALISTS BEYOND THE TEACHER-LIBRARIAN WERE ABLE TO GET THEIR FOOT IN THE CLASSROOM DOOR.

What does the ideal reading class look like in light of 21st-century skills and the demand for higher levels of college and career readiness? One district set out to develop and implement such a course.

BACKGROUND

North East ISD (Independent School District) is a recognized school district in San Antonio, TX, with an approximate enrollment of 61,000 students. North East has 13 middle schools with grades 6–8. While the state of Texas only requires a reading course in sixth grade, North East is deeply committed to student success and knows that students who are proficient readers are also most often proficient students. As a result, traditional reading courses were required for all three middle school grades. These courses primarily consisted of teaching comprehension skills and literary elements through reading fiction and novel studies.

Students were enrolled in the reading courses regardless of their ability level. Many students, teachers, and parents were concerned about the instructional rigor of this middle school reading experience. This was especially true of gifted and talented (GT) and pre-AP students for whom reading comprehension was not a problem. The district acknowledged this concern and turned to a team of curriculum specialists to develop an alternative course for those students who were ready for a more rigorous curriculum.

COURSE DESIGN

A team of literacy specialists with GT certification created the initial design of each Advanced Contemporary Literacy course. Using their knowledge and experience of GT principles, 21st-century skills, and instructional best practices, they outlined the structure for the new course at each grade level. The specialists then generated overarching themes, essential questions, and reading and discussion formats. They also enumerated academic vocabulary and suggested readings from the state-adopted textbooks and supplemental readers.

The initial work resulted in the following brief course description:

Each course is rigorous and designed for students possessing advanced reading skills. Students will apply their reading skill in problem-solving, conducting increasingly sophisticated research, and completing project-based assignments. Instructional strategies will include Socratic seminars and student portfolios to encourage higher level thinking and discussion of contemporary cross-curricular topics.

The components required in each course are the use of nonfiction material, Socratic circles, research, and presentation. The research component is a spiraling feature that begins by teaching the Big6 research process to sixth-graders during the first 9 weeks, then each following unit suggests appropriate research models from David Loertscher's *Ban Those Bird Units* (2005). Several of these models are repeated from one grade to the next, with new models added each year to accommodate increasingly complex research issues.

sharon swarner

SCHOOL LIBRARY 2.0: FROM THE FIELD

In addition to research, each grade level has a primary presentation emphasis. Sixth grade focuses on oral presentations and written expression. Seventh grade emphasizes, "reading" contemporary media, and eighth grade teaches students to formulate arguments and present their positions in a debate format. Underlying all these elements is the concept of student choice. The themes and essential questions are broad enough to spark students' interest, and the research models are expansive enough to allow student interest to shape both the topic selection and the research model used.

All of this foundational work set the stage for a larger group of practitioners to refine the course. A large committee was assembled and subdivided into grade-level subcommittees, each consisting of at least one literacy specialist, reading teacher, instructional dean, technology specialist, and teacher-librarian. Using the framework materials provided by the literacy specialists, the subcommittees selected the themes and refined the essential questions that addressed the overarching theme. They also set the order in which the themes would be introduced by taking into consideration the scope and sequence of other subject areas. The subcommittees further selected the appropriate research models, suggested critical reading strategies, academic vocabulary, and additional reading selections that would advance the exploration of the theme. All of these refinements were the result of deep, focused collegial conversation.

For example, the theme for the first 9-week session of the eighth-grade course is "Shades of Truth." The essential questions include "Are there certain truths that can be considered universal or absolute?" and "How does one's perspective shape or alter truth?" The suggested readings that support these questions are selections from the Koran and the *Tao Te Ching* (Tzu, 2006), as well as pieces from the regular textbook, such as *The Last Seven Months of Anne Frank* (Lindwer, 1992). Academic vocabulary such as *bias*, *propaganda*, and *persuasive techniques* are suggested as are critical reading strategies like "identification of speakers purpose and tone" as well as "noting the writer's word choice" to hone an awareness of diction. In addition, research models are suggested to correlate with the theme. In this particular instance, the suggested models are *Background to Question Model* or *Take a Position*. During the course of the rest of the year, these same eighth-graders will explore themes of social justice, freedom and responsibility, and the pursuit of happiness, all the while reading complex nonfiction passages, wrestling with complicated questions, and researching topics that relate these issues to their own lives.

IMPLICATIONS FOR THE LIBRARY PROGRAM

The Advanced Contemporary Literacy (ACL) course is a model in curriculum development. Teacher-librarians were invited to the table at the inception of the course because research is an integral component of the curriculum requiring extensive instruction and use of research models and skills. Teacher-librarians take the lead in teaching the kind of research skills and processes that allow students to engage in high levels of thinking and communicating. In addition, the marked increase in student-driven research has given a clear focus to our collection development of both print and electronic resources.

Teacher-librarians are also invited to professional development days that are set aside throughout the year. These days provided time to talk about successes and concerns. There is also afforded time to plan the following 9 weeks and to work with specialists on specific aspects of the curriculum. These collaborative planning sessions were important to the first-year success of the ACL classes. This open communication and collaboration continued on the campuses offering teacher-librarians the opportunity to be cocreators and coteachers in a truly integrated reading curriculum.

AND WHAT ABOUT THE RESULTS?

We have anecdotal evidence from teachers, students, and parents, even though, more solid evidence would be better. We saw teachers who became more energized throughout the school year as they got further into the curriculum and the instructional methods. Part of the professional development days included a time to share their successes. They explained student projects. One particularly interesting project featured seventh-graders at one campus who had their class think about local challenges facing the community. They "researched and created a variety of media, including logos, pamphlets, commercials, billboards, web sites, inventions, and more." Following their research, the students "implemented their plans and successfully raised funds for their cause." Teachers also shared activities such as student-created public service announcements (PSA). These PSAs were a direct result of themes they had studied in the ACL course. Sharing these types of class activities created a sense of camaraderie and a challenge to keep the class growing and working at a high level. We were not able to track the state assessment test scores for our ACL students this year. We hope to do that again next year.

Upon reflection, there remain challenges given such a drastic change in curriculum and collaboration. However, we are marching along with the confidence that teacher-librarians and high-think learning experiences pay big dividends.

REFERENCES

Lindwer, W. (1992). *The last seven months of Anne Frank.* New York: Anchor.

Loertscher, D. (2005). *Ban those bird units! 15 models for teaching and learning in information-rich and technology-rich environments.* Westport, CT: Libraries Unlimited.

Tzu, L. (2006). *Tao Te Ching: A New English Version.* Trans. S. Mitchell. New York: Harper Perennial.

Sharon Swarner has been the Library Technology Coordinator for North East ISD since 2001 where she assists librarians and collaborates with curriculum specialists. She may be reached at sswarn@neisd.net.

Part VI:

Assessment in the Learning Commons

The final section in this compilation addresses the obvious need to assess the impact of Learning Commons programs and practices on teaching and learning. Obviously, a full range of assessment practices needs to be considered in any whole look at impact, but the three authors make some headway in the role of self-assessment, documentation of impact, and the role of meta-cognition in measuring success and making improvements.

For even more information, readers are referred to the sister publication to this volume for a larger view with many planning sheets and checklists: *Building the Learning Commons: A Guide for School Administrators and Learning Leadership* Teams by Koechlin, Rosenfeld and Loertscher (Hi Willow, 2010).

It is certain that the impact of the Learning Commons will not appear on tests of factual knowledge. In this section and elsewhere, the authors are looking at the power of technology, engagement, and other practices on 21st Century skills in combination with the deep understanding of content knowledge. The central point of assessment here and in the other two publications mentioned in the introduction are that a broad spectrum of assessments is much preferable to a single high stake test. Teaching to a single test is not a part of any Learning Commons strategy. Instead, the message here is to broaden and triangulate evidence from a wide range of both formative and summative findings as a true barometer of its health in an educational environment.

FEATURE ARTICLE

Creating Personal Learning through Self-Assessment

Self-assessment is a habit of mind that engages one in metacognition and reflection.

JEAN DONHAM

> "Habits are behaviors we exhibit reliably on appropriate occasions and they are smoothly triggered without painstaking attention" (Costa & Kallick, 2000, p. viii).

The American Association of School Librarians' (AASL) *Standards for 21st Century Learners* (2009) provide a set of outcomes intended to develop lifelong learners; these standards delineate skills, knowledge, responsibilities, and dispositions essential for independent learning. In addition, they call for students to develop the capacity for self-assessment in the context of their information seeking.

Self-assessment is a habit of mind that engages one in metacognition and reflection. Such reflective behavior calls for intentionally reviewing our own performance based on criteria that we have internalized. We can think of it as metacognition with an evaluative bent. When we are metacognitive, we are aware of what we know and what we do not know. We are able to plan a strategy for producing what information is needed, to be aware of our own strategies as we problem-solve, and to reflect and evaluate our own productiveness.

These behaviors constitute the mindfulness necessary to be self-reliant learners for life. In order for these behaviors to become habits, we must intentionally build these capacities and automate these behaviors by continuous practice. Teacher and teacher-librarians have the important role of building the habit of self-assessment in students so they can indeed become self-directed learners who know for themselves what they don't know and can plan to self-correct.

The chains of habit are generally too small to be felt until they are too strong to be broken. ~Samuel Johnson

Teacher-librarians support the development of habits of self-assessment by consistently expecting students to reflect on their work and by providing them structures and criteria to guide the process. As a result of practice, they can become accustomed to self-assessment as a way of working.

CONTEXT

The AASL *Standards for 21st Century Learners* acknowledges the complexity of information literacy as more than a set of skills; they are grounded in recognition that information literacy

- emerges from a disposition of inquiry—what do I want to know about?
- requires a skill set—how can I be most efficient at finding the information I need?
- embraces a disposition of critical thinking—what is the authority of the source of information?
- assumes an approach that is analytical and/or interpretive—what meaning can I make of these findings?
- calls for an understanding that each discipline poses its own genre of questions and requires familiarity with a subject and the nature of its inquiry.
- establishes a stance of self-assessment—how do I know that I have the information I need?
- is for life—I will always need to be learning.

IMPORTANCE

We are living in an "other-directed" era in education, according to Arthur Costa and Bena Kallick (2004). That is, learning is measured in scores on tests externally designed and scored. Such externalization of decisions, about teaching and learning, stands in contrast to the foundation of lifelong learning.

Lifelong learning is a personal, self-directed process that begins with self-generated questions and curiosity, includes self-directed inquiry and exploration, and concludes with self-measured success. The imposed other-directed experiences in schools will not be adequate to prepare students to continue to learn effectively beyond school.

Students recognize this sense of other-directedness. For them, their teacher directs their learning. Speaking of other-directed learning, Michael Gordon (1999) opines in *Education for Health*:

> An auditor checks on the banker's ledgers. A building inspector signs off on the engineer's drawings. But who signs off on the physician's orders? Typically no one. Physicians [and many other professionals] operate under the social contract for legitimate autonomy. In exchange for this privilege they are expected to self-regulate; to keep abreast of fast-moving advances that affect the health of patients and communities. Meeting these expectations depends not only on their willingness but also on their ability to assess their professional knowledge and skills and to act constructively on these assessments throughout their long careers (p. 167).

Gordon continues in this essay to make an argument for developing skills of self-assessment as crucial to effective professional performance in a world of change. In addition, he maligns medical education for its emphasis on other-directed (and he says demeaning) forms of assessment.

Engagement

Clearly, the ability of physicians to self-assess can be a matter of life and death. It is important to believe that even with the other-directed and externally imposed agenda in K-12 education today, individual teachers and individual teacher-librarians can develop habits of self-assessment in their students—habits that will help them be self-reliant learners for life.

Consider the effect of self-assessment on student engagement. Munns and Woodward (2006) discuss the discourses of power in the classroom; they suggest that in many traditional settings, students feel powerless in the classroom. They identify five discourses of power that affect student's self-perception as learners. These are

Knowledge—Why do I have to learn this?—resulting in disinterest in the task at hand.

Ability—I don't believe I can do this—resulting in low aspirations.

Control—feelings that the student has little or no control over what he/she will do.

Place—"I am just a kid from. . ." particularly debilitating for students from lower SES settings, ELL students, and students with special needs—they feel devalued as learners assuming that whether they learn or not is unimportant.

Voice—lack of say over learning as the teacher controls the content and sits in judgment of the performance.

Self-assessment may be a way to give students a sense of ownership of their learning—and thereby responsibility for it.

Too often students complete an assignment, get the grade, and move on without looking back. When we allow this, we miss out on one of the most important reasons for assessment—to use it as a learning opportunity. Students need opportunities to review their work as it progresses through formative assessment and then to look back on finished work in order to look forward to the next task with strategies for improvement. This requires self-assessment and reflection to be regular components of each lesson or session in the school library throughout a unit or project. Their final assessment of their own work affords the opportunity to think about the "so what?" of the learning experience—to answer the question, "I got the assignment done—so what? What did I learn about the content and what can I learn from the process I used to accomplish the task?" Without intentionally engaging in the self-examination inherent in these questions, the learning should come from each experience may be lost so that there is less likelihood that the next experience will be better.

Excellence

This brings into focus an important disposition inherent in self-assessment—a disposition toward excellence. Such a disposition needs to be oriented toward the long-term—not only toward satisfying the teacher. Self-assessment affords students the opportunity to cultivate both the strength and humility to continue to learn. Likewise, though, throughout life, the ability to self-assess will be the mark of a lifelong learner in any walk of life—the disposition to pose and then the ability to answer fundamental questions related to information:

Do I have enough information?

Do I have the right information?

Have I interpreted and applied the information appropriately?

How will I improve my use of information next time?

To believe in self-assessment as essential to lifelong learning is to believe students can and should reflect upon and assess their own performance. We must believe in the importance of developing in students the disposition that such self-reflection is productive, and we must develop in them the skills and strategies to engage in self-assessment.

Self-assessment means developing internal standards and comparing performance to those standards. While teachers can assess students' products as manifestations of their learning, it is the student who can assess his/her thinking, attitudes, motivations, and learning processes. Only the student can know how he/she chose to be finished seeking information. Only the student can assess how he/she reconciled differences among sources. Only the student can assess the ways in which information was interpreted to serve his/her purpose. Only the student can assess the degree of willingness to live with ambiguity or uncertainty versus the desire for certainty or finality.

As in Figure 1, self-assessment requires looking in three directions: back at completed work, down at present work to determine next steps, and forward to the future to apply to the next learning opportunity what has been learned. Self-assessment is essential to learning as the learner reflects upon what is known, what remains to be known, and what will be required to fill the gap between them.

For students to engage in self-assessment in these three directions, they are by default engaging in both formative and summative self-assessment, i.e., self-assessments that occur as they are in the inquiry process to help them progress and summative assessment that looks back at completed work to determine what went well and what should be different next time.

Alverno College in Milwaukee, WI, has been working at developing a strong self-assessment program for decades. Their work is based on the belief that students must leave school able to assess their own work as independent learners. Without such ability, how will they know what they must next learn? The administrators at Alverno College have identified four components of skills inherent in self-assessment:

- Observing: This requires the student to observe his/her own work using specific criteria.
- Analyzing: Requires students to use criteria to examine work; analysis typically means taking the whole and breaking it into meaningful parts to be examined closely.
- Judging: Means students must arrive at a conclusion about the quality of their work as a result of the comparison between a standard for performance and one's own performance.
- Planning calls for thinking ahead as to how one uses the outcome of a judgment to improve performance next time (Alverno College, 2001).

These are the very acts students perform when they self assess—observing their own performances, analyzing their work according to internalized criteria, applying those criteria to judge quality, and using their self-observations to plan for their next performance.

Figure 2. Search Process Model in Student Language (Donham, J. 2001).

DISTRICT INFORMATION SEARCH PROCESS MODEL	"KID" LANGUAGE
Select and define the topic	What do I want to study about <topic>?
Write and categorize known information	What do I already know?
Gather, evaluate and select materials	Let's read and find out.
Understand materials	What does that mean?
Select and record information	Let's write it down.
Organize notes and discoveries	Does it make sense?
Prepare and present	Let's put it all together. Let's share what we have learned.
Assess our process and product	What did I learn? What did I do well? What will I do differently next time?

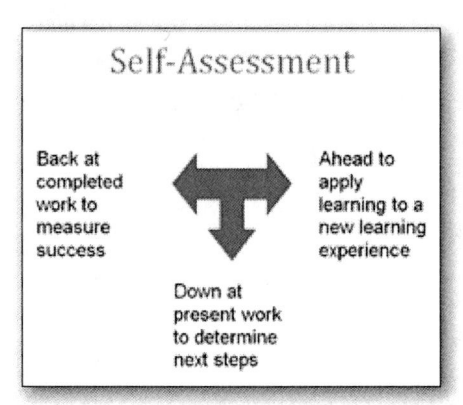

Figure 1. Self assessment is multi-directional.

STRATEGIES FOR DEVELOPING SELF-ASSESSMENT HABITS AND SKILLS

Teachers and teacher-librarians can integrate self-assessment into information literacy programs, especially when working collaboratively. To ensure that self-assessment happens opportunities for both formative and summative assessment need to be built into the design of the lessons and activities from the beginning. It is too easy to arrive at the end of a unit or project and "run out of time" unless there is intentional planning for self-assessment. A number of strategies serve as options for integrating self-assessment into the context of teaching and learning.

A. *Direct instruction.* Explicit and intentional explanation of the self-assessment process and guided practice help students develop first an understanding of self-assessment and then the necessary skills of observation, analysis, judgment, and planning. In this framework, Gordon (1999) recommends three steps to an effective self-assessment program:

- Students must independently evaluate their own competencies using an evaluation tool.
- Teachers and/or teacher-librarians, using the same evaluation tool, also independently evaluate how the students perform.
- Students and teachers compute their separate evaluations, and compare and reconcile any differences found.

The criteria and/or evaluative tool can be generated by the teacher or collaboratively by the teacher and the class by examining the assignment and determining the criteria suggested by that assignment. Through direct instruction the teacher and teacher-librarian teach students how to use assessment tools and strategies, set expectations that students will use self-assessment tools, and provide feedback to students in their assessment processes.

a. *Scaffolding.* There are various techniques for scaffolding the self-assessment process.

The Information Process Model. By internalizing a model of the information search process, students acquire a framework upon which to reflect and judge their progress in information work independently. Wolf (2003) examined the use of an information search process model to scaffold metacog-

Figure 3. Search Report Process Guide

Topic: _____

Research is a recursive process in which we search, revise our search strategy, then search again, revise our strategy, etc. Fill in the grid below for each search to track process.

Date of search	Tool used (EBSCOhost MAS, for example)	Search terms (exactly what I typed)	Number of results	Did I limit my search in any way? If so, how?	Assessment of results (Were results useful? If not, why not?)

How would I rate my success at finding the information I wanted? _____

What problem(s) did I encounter? _____

What will I do differently next time I begin a search for information? _____

Reflection Question: How does a search process model influence the work that your students do in your school?

"By encountering a consistent model across grade levels and disciplines, students can be intentionally conscious of where they are in the research process, what they should be expecting of themselves at each stage, and how they can measure their progress. This suggests that teacher-librarians enlist the support of teachers throughout their schools to adopt as 'official' one process model that will provide consistent language and a consistent framework for students' self monitoring of their information search process."

---

nition and found that students' use of the "Big Six" model afforded students a scaffold useful for self-monitoring their progress. Popular examples of research models include those developed by Kuhlthau (2004), Pappas (Zimmerman, Pappas, & Tepe, 2002), as well as Eisenberg and Berkowitz (1990). Findings in one of the *Library Power* studies demonstrated that school-wide adoption of a single information search process in an elementary school provided the appropriate lexicon for monitoring and communicating about research progress (Donham, 2001). In this case, the process model was taught to children using "child-friendly" language to define each stage of the process and posters of the model in that language appeared in classrooms throughout the school. (See Figure 2). By encountering a consistent model across grade levels and disciplines, students can be intentionally conscious of where they are in the research process, what they should be expecting of themselves at each stage, and how they can measure their progress. This suggests that teacher-librari-

ans enlist the support of teachers throughout their schools to adopt as "official" one process model that will provide consistent language and a consistent framework for students' self-monitoring of their information search process.

b. *Reflection Logs.* Teacher-librarians can require students to periodically respond to prompts as they proceed through a research project. These may be simple prompts like:

What frustrations did you encounter today?

What new question has arisen in your research today?

What was an important discovery you made today?

Consistent opportunities to reflect on a "chunk" of progress can lead to habit formation.

c. *Search Log.* Providing a student with a framework for monitoring their search can serve as the scaffold for self-monitoring (See Figure 3). It is important that students review the log and reflect on what the log

> "Simply completing and submitting the log does not demonstrate to students the benefit of reflection - instead, it then becomes an external measure of their work by the teacher, but not valued for what students can teach themselves."

Figure 4. Multi-column note-taking

Source of information:
_____
_____

What I read	What I think about it
Idea 1.	
Idea 2.	
Idea 3.	

What next:
- Find more information about _____
- Compare this information with_____
- Apply these ideas to this sub-question_____

teaches them about their process and how they would plan to search differently in the future. Simply completing and submitting the log does not demonstrate to students the benefit of reflection—instead, it then becomes an external measure of their work by the teacher, but not valued for what students can teach themselves. After all, self assessment and lifelong learning require that students learn how to analyze and self-correct their performances.

d. *Multi-column Note-taking.* Students record information in the wide column and in the narrower one, reflections about the information—how they will use it, whether it is incomplete and will require further research, etc. This technique again creates a framework for students to be explicit in studying their progress through self-assessment. Figure 4 illustrates a template that students might use in note-taking and features the "what next" questions at the bottom of the template.

e. *3-2-1 Form.* Often the summative assessment falls by the wayside under the pressure to move on to the next unit or

Figure 5. 3-2-1 Summative Self-Assessment

3 tips about information searching that I learned

1.
2.
3.

2 aspects of my inquiry work that make me proud

1.
2.

1 mistake I made that I will try not to make again

1.

next activity. Or, the summative assessment involves only assessing students' content learning. A simple and quick way to provide a framework for students to self-assess at the end of an experience is to provide a 3-2-1 form (see Figure 5). The prompts can be composed to match the assignment and expectations of the student, or they can be somewhat generic—as long as they focus on the inquiry process.

f. *I-Search.* Ken Macrorie's I-search model for writing research papers affords the opportunity for students to attend to not only to their findings but also to their methods (Macrorie, 1988). By writing in the first person and describing their research process as they reveal their findings, students are engaging in a metacognitive writing process. Simply speaking, the structure of the I-search paper begins with opportunities for background building in a broad topic or theme and the goal is to arrive at a research question—what does this information leave me wondering? Next, students develop a search plan that identifies how they will gather information: by reading books, magazines, newspapers, reference materials; by watching videos; by interviewing people or conducting surveys; or by carrying out experiments, doing simulations, or going on field trips. Then, students follow their search plans and gather information. They also analyze and synthesize information to construct knowledge. Finally, students draft, revise, edit, and publish an I-Search Report. The I-Search Report includes the following components: *My Search Questions, My Search Process, What I Learned, What This Means to Me,* and *References*. Because students are not only writing their findings but also describing their process, the I-search structure builds in opportunities for self-assessment and metacognition.

C. *Modeling.* Rubrics and checklists provide a structure for self-assessment. They become more empowering when students contribute to their development. By participating in the design of the rubric, students can gain more independence in learning how to self-assess. In a study of third and fourth grade student writing, Andrade, Du, and Wang (2008) found that using a rubric to self-assess was positively related to the quality of writing. This finding is consistent with other studies that reported positive effects on academic performance when students engage in self-assessment (e.g., Ross, Hogaboam-Gray, & Rolheiser, 2002 and Andrade & Boulay, 2003).

Beginning with the assignment, working together as a class to construct the rubric allows opportunities for feelings of empowerment, but also for developing a skill that will help students assess when no one is there to provide the prepared rubric for them. Such a process constitutes modeling when the teacher explicitly verbalizes the process of using the assignment to build the rubric. Figure 6 provides a framework for developing a rubric collaboratively. The first few cells are completed as examples of the kind of description students might generate with guidance from the teacher-librarian. Students and teacher can develop the text to describe levels of performance to be filled into each cell. Effective modeling requires thinking aloud or "labeling the steps" of a process one is modeling. By saying aloud the questions we are asking ourselves about our performance, we render visible our thought processes and provide a model for students. Like the development of any habit, students must have multiple opportunities to hear us thinking aloud in order to recognize it as a habit to be developed.

D. *Peer Assessment.* Learning is a social process and peer assessment calls for students to engage in conversation that forces them to articulate their analysis and judgment of the work they do. This process mirrors the way we often assess our own work outside the structure of school. We take our work to someone and request his or her opinion of it and we engage in this process best when we know what to ask for. In a high school classroom, for example, this process can be practiced by having students confer with a "critical friend." These peer consultations can occur at various stages of the process, but they require structure and focus. So, early in the research process, students can be partnered to serve as one another's "critical friend." An initial conversation may be structured around the research question. Novice pairs will need to be given somewhat scripted questions:

What are you interested in?

> "Learning is a social process and peer assessment calls for students to engage in conversation that forces them to articulate their analysis and judgment of the work they do."

Figure 6. Student –Developed Rubric			
Research Stage	**Expert**	**Proficient**	**Novice**
Choosing a topic	I chose a topic that I was personally curious about	I chose a topic that my teacher suggested that interested me	I chose a topic that I thought would be good for this assignment
Exploring my topic	I looked around at Internet sites and library books to get ideas about my topic	The teacher librarian helped me find some information	I found one website or one book about my topic
Focusing a my topic into a research question			
Locating information on my topic			
Organizing information and discoveries			
Communicating my results			

> "In the long run, by internalizing a model of the information search process, the critical peer will be able to generate the questions for consultations independently."

> "The power of this process is not so much in the assistance students receive from each other, but instead in the necessity for them to articulate aloud their own perception of their progress."

What do you already know?
What do you wonder?
What might be your research questions?
How do you expect to begin to look for information on these questions?

Later in the process a conversation might be guided by these questions:

What is your research question? Has it changed?
What difficulties have you encountered in finding information?

And, finally, the conversation may be:

What went well with this project?
What could have gone better?
How could you make the research experience better next time?

The hope is that students can offer suggestions to each other and individuals have the opportunity to self-assess their own progress as they articulate their research progress to someone else.

In the long run, by internalizing a model of the information search process, the critical peer will be able to generate the questions for such consultations independently.

Younger students can engage in such conversation as well—either with each other or with their teacher or teacher-librarian. The power of this process is not so much in the assistance students receive from each other, but instead in the necessity for them to articulate aloud their own perception of their progress.

Talking is a way of learning because it requires the learner to construct in language descriptions of their own work. In this way, learners become more aware of:

- what they learn
- how they learn
- what helps them learn.

CONTINUOUS LEARNING AND ASSESSMENT

The long-range goal is to develop in students the ability to be continuously learning—allowing every learning experience to inform the next one. Self-assessment means asking, "What did I learn? What did I do well? What will I do differently next time?" Consistently posing these three questions can help self-assessment to become a habit—a behavior as Costa and Kallick (2000) says, "we exhibit reliably on appropriate occasions and they are smoothly triggered without painstaking attention." By developing habits of self-assessment we encourage our students to evolve from dependence on others by cultivating the habit of continuous personal learning—an essential life habit in a world of change. Self-assessment turns learning activities into opportunities to increasingly think more deeply, more creatively, and more critically.

REFERENCES

Alverno College. (2001). *Student assessment as learning.* Retrieved June 16, 2009 from http://depts.alverno.edu/saal/selfassess.html.

American Association of School Librarians. (2009). *Standards for the 21st century learner.* Chicago: American Library Association.

Andrade, H. & Boulay, B. (2003). The role of rubric-referenced self-assessment in learning to write. *Journal of Educational Research,* 97 (1), 21-34.

Andrade, H. L., Du, Y., & Wang, X. (2008). Putting rubrics to the test: The effect of a model, criteria generation and rubric-referenced self-assessment on elementary school students' writing. *Educational Measurement: Issues and Practices,* 27 (2), 3-13.

Costa, A. & Kallick, B. (2000). *Discovering & exploring habits of mind.* Alexandria, VA: Association for Supervision and Curriculum Development.

Costa, A. & Kallick, B. (2004). Launching self-directed learners, *Educational Leadership,* 62 (1), 51-55.

Donham, J. (2001). The importance of a model. In Donham, J., Bishop, K., Kuhlthau, C.C., & Oberg, D. *Inquiry-based learning: Lessons from Library Power.* Worthington OH: Linworth Publishing.

Eisenberg M. B. & Berkowitz, R. E. (1990). *Information problem-solving: The big six skills approach to library and information skills instruction.* Norwood, NJ: Ablex.

Gordon, M. (1999). Commentary: Self-assessment skills are essential, *Education for Health,* 12(2), 167-168.

Kuhlthau, C. C. (2004*). Seeking meaning: A process approach to library and information services, second edition.* Westport, CT: Libraries Unlimited.

Macrorie, K. (1988*). The i-search paper.* Portsmouth, NH: Boynton/Cook Publishers.

Munns, G. & Woodward, A. (2006). Student engagement and student self-assessment: The REAL framework. *Assessment in Education.* 13(2), 193-213.

Ross, J. A., Hogaboam-Gray, A., & Rolheiser, C. (2002). Student self-evaluation in grade 5-6 mathematics effects on problem-solving achievement. *Educational Assessment* 8(1), 343-59.

Wolf, S. Brush, T., & Saye, J. (2003). The big six information skills as a metacognitive scaffold: A case study. *School Library Media Research,* Volume 6. Retrieved on June 16, 2009 from http://www.aasl.org/ala/mgrps/divs/aasl/aaslpubsandjournals/slmrb/slmrcontents/contents.cfm.

Zimmerman, M., Pappas, M., & Tepe, A. (2002). Pappas and Tepe's pathways to knowledge model. *School Library Media Activities Monthly* 19 (3), 24-27.

Jean Donham, PhD, is a faculty member and associate professor in library science at the University of Northern Iowa. She is the author of *Enhancing Teaching and Learning: A Leadership Guide for School Library Media Specialists.* She is co-editor of *School Library Media Research* journal and may be contacted at jean.donham@uni.edu.

Articles in *TL* are blind refereed by members of the advisory board. This article was submitted in August 2009 and accepted in November 2009.

"Self-assessment turns learning activities into opportunities to increasingly think more deeply, more creatively, and more critically."

TL EXTRA

Our Instruction DOES Matter! Data Collected From Students' Works Cited Speaks Volumes

As librarians, it is sometimes difficult to gauge in hard numbers the influence our instruction has on student end results

SARA POINIER AND JENNIFER ALEVY

Each year we set goals for our library program, striving to reflect on and steadily improve student achievement.

Last year one of our program goals was to complete a common assessment for one class. After learning about the idea of using students' works cited as a way to reflect on the value of our involvement with classes at the summer 2008 IASL (International Association of School Librarians) Conference, we attempted this data collection method in our high school.

In this endeavor we learned not only about our students' learning but also what an effect our instructional role has on student achievement.

COLLECTING DATA

In fall 2008, we arranged with our health teachers to provide instruction on both library resources and the works cited page document for a class project the students undertake each semester.

Over the course of several semesters we had been steadily increasing our collaboration with the health department in our school after teachers expressed dissatisfaction with the quality and integrity of their students' work. Given librarian instruction about reliable resources, ethical use of resources, and plagiarism teachers had reported improvement in student project quality. To further quantify this success, we planned to collect the students' works cited pages in order to discern their use of reliable information and the accuracy of the works cited format.

Because we wanted to collect comparative data from a group of students who received no librarian instruction, we also enlisted the help of one of our science teachers whose class happened to be conducting library research at the same time. She planned to collect works cited pages from her students, but made no plans for librarian instruction about resources or works cited.

Our Process

After providing each health class with a short introduction to reliable resources and instruction on how to create citations and the works cited page, the students got to work. Several days later, students were expected to turn in a works cited page to both their teacher and to us the teacher-librarians. Using Google Apps (our school district has a subscription) we tabulated our results. We literally counted the number of reliable and unreliable resources on each works cited page. We deemed reliable the following: books from the library, on-line database articles, and web sites we had recommended to students. Additionally, we rated the works cited on a 0/1/2 scale for the following formatting issues: title, alphabetizing, double spacing, and hanging indentation. Student work received a 0 if they did not complete this aspect correctly, a 1 if they inconsistently demonstrated mastery, and a 2 if they consistently applied this criteria.

The Result: A Tale of Two Classes

The results of this study were telling. The Astronomy class, which came into the library and worked without instruction, used mostly what we would consider unreliable web resources—mainly web sites not affiliated with any reputable expert group on the topic. Only one library-provided resource, online or in print, was used by any of these students. The works cited pages turned in by this class were more of a collection of URLs than documentation of sources (see all results in Table 1).

Conversely, the health students submitted works cited pages that used 81% of the library recommended resources–a dramatic difference. In addition, the health students turned in works cited pages that showed a decent effort at being correctly formatted (see Table 2 for complete information). Thus, we drew the following conclusions: collaborative planning and instruction by the teacher-librarian leads to both an increase in the use of reliable resources and an improved works cited product as compared to a class that has simply "used the library."

REVISION COUNTS

Not content to see that our health students still had works cited pages riddled with formatting errors, we sought to improve our instruction for the second semester. We revised the handout we gave to students and found a better way to highlight what needed to be accomplished.

What we found after using the new and improved handout and giving a more instructional focus to the works cited page itself was that improvement was made on every measure during second semester from 4% more students correctly using a hanging indent to 26% more students including and appropriately placing a title on their works cited (see Table 3).

As librarians, it is sometimes difficult to gauge in hard numbers the influence our instruction has on student end results, yet collecting and scoring our students' works cited pages demonstrated that our role as collaborative, reflective practitioners makes a difference.

Sara Poinier and Jennifer Alevy are teacher-librarians at Horizon High School in Thornton, Colorado. Poinier can be contacted at sara.poinier@adams12.org and Jennifer Alevy can be contacted at jennifer.r.alevy@adams12.org.

Table 1: Fall Semester Astronomy Class Control Group, Grades 11/12

Number of Classes	1
Number of Students	32
Number of Library Database Articles Used	0 (0%)
Number of Library Books Used	1 (3%)
Number of Recommended Web Sites	0 (0%)
Number of Non-Recommended Web Sites Used	28 (97%)
Total number of Library Recommended Resources Used	1 (3% of all resources used)
Average Score: Alphabetize	0 (out of 2)
Average Score: Double Space	0 (out of 2)
Average Score: Hanging Indent	0 (out of 2)
Average Score: Title	1.25 (out of 2)

Table 2: Fall Semester Health Classes, Grades 9-12

Number of Classes	5
Total number students	178
Number of Library Database Articles Used	150 (52%)
Number of Library Books Used	52 (18%)
Number of Recommended Web Sites Used	32 (11%)
Number of Non-recommended Web Sites Used	54 (19%)
Total Number of Library Recommended Resources Used	234 (81% of all resources used)
Average Score: Alphabetize	1.39 (out of 2)
Average Score: Double Space	1.01 (out of 2)
Average Score: Hanging Indent	1.17 (out of 2)
Average Score: Title	1.26 (out of 2)

Table 3: Spring Semester Health Classes, Grades 9-12

Number of Classes	4
Total Number of Students	134
Number of Library Database Articles Used	109 (64%)
Number of Library Books Used	2 (.01%)
Number of Recommended Web Sites Used	29 (17%)
Number of Non-Recommended Sites Used	31 (18%)
Total Number of Library Recommended Resources	140 (82% of all resources used)
Average Score: Alphabetize	1.53 (out of 2)
Average Score: Double Space	1.43 (out of 2)
Average Score: Hanging Indent	1.25 (out of 2)
Average Score: Title	1.78 (out of 2)

FEATURE ARTICLE

The Big Think:
Reflecting, Reacting, and Realizing Improved Learning

When we put our heads together with classroom teachers, we want one plus one to equal three!

CAROL KOECHLIN AND SANDI ZWAAN

The game has ended, and the scores have been tallied. What were the results? Are we satisfied? Would we have liked something better?

What do winning teams do when they are not satisfied with their performances? They pick themselves up, rewind the tapes, review, and observe. The coaches and the players analyze their successes and look for the possible causes of their less than stellar plays. Even winning teams review their play and begin to plan the strategy for the next game, building on the positives and attacking the weaknesses with renewed energy and commitment.

If as classroom teachers, teacher-librarians, and learners we do not take a similar action, if we continue using the same strategies and processes we have always used then we can expect only a repetition of the same outcomes. So how do we accomplish the 'post game' review? We do not have the luxury of 'days off between games'. In education there is a need for a continual stream of assessment of the learning; not just the knowledge and understanding of content, but also the effectiveness of the strategies and processes used to achieve that learning. What we need is a streamlined, easy to apply approach that both teachers and learners can use effectively and efficiently as our units draw to a close and we begin to plan for the next activity. To get better as learners we must apply ongoing metacognitive assessment strategies that appraise what we know and how we learned it and inspire us to take action.

Many of our readers are familiar with The Think Models (Loertscher, Koechlin, & Zwaan, 2007) we created a few years ago to replace the common low-level bird units that plagued school libraries. The models offer a better way to 'play the game' because they provide stages of high think inquiry, information processing, and opportunity to build on the knowledge and expertise of others. During the process learners take on more and more responsibility for their own learning as they utilize the best resources, technologies and strategies to their advantage. The classroom teacher and teacher-librarian's role is to ignite interest, guide and coach learners, and provide ongoing metacognitive assessment throughout the learning experience thus building essential learning to learn skills. To ensure that learners are aware of the content and skills they have gained in the unit, each of the Think Models also wraps up the experience with a Big Think so everyone is cognizant of what they have learned and how they learned it.

Over the last couple of years, as we have coached teacher-librarians and teachers through these models and the design of High Think inquiry, we observed a need to expand our work on the Big Think culmination activity. This deliberate metacognitive experience has even more value than we originally thought. It has the potential to change everything!

Why do we give students research projects? What do the students gain? How do we know they have benefited? How do students know if they have gained anything? What do teachers learn from these assignments? Do we have evidence that our inquiry assignments contribute to school improvement? Are we keeping pace with the needs of learners today?

When we asked these questions in workshops and with individual students we were disappointed with the answers we received, consequently we researched, rethought, and expanded our concept of ending formal units of study and research assignments in a big think. The outcome is our book called *The Big Think: 9 metacognitive strategies that make the unit end just the beginning of learning* (Loertscher, Koechlin, & Zwaan, 2009), which develops nine metacognitive strategies that can be used with any ability and grade level and any subject to ensure that everyone—students and teachers—not only gain from the main experience but also are aware of what they now know, how they learned it, and how they can improve the learning. Like athletic coaches, we want our team to get better and better every "game" we play in our drive toward excellence.

We propose an idea so simple yet so rewarding it really is worth the investment. By engaging in the Big Think, as teacher-librarians we can triple the benefits of our efforts. With these three important 'returns' on our investment we can influence teaching and learning on a school wide basis.

When we put our heads together with classroom teachers, we want one plus one to equal three! Our focus as we watch the rerun of the learning experience as coaches and learners together will be on three main things that happened during our game together: analysis of learning how to learn, how we taught them to learn with our team players, and how our game strategy affects school improvement.

RETURN #1: LEARNING TO LEARN WITH OUR TEAM PLAYERS

Instead of just setting aside individual learning at the traditional end of the unit and moving on to the next topic, the Big Think enables learners to build on each other's expertise and pool their collective knowledge to do some deep thinking and working with this body of new ideas and information. This collaborative knowledge building does not mean that ideas are distilled or meshed together to produce a consensus product. Instead it means that individual knowledge is considered, analyzed, and worked by groups to build a new richer understanding that can only occur once they can see the big picture. When learners are provided this opportunity, content knowledge is broadened and deepened, fresh perspective is gained, and lasting understanding takes hold.

Collaborative knowledge building is a desired outcome of working, playing, and learning today but it does not just happen. Educators need to develop the knowledge and skills that are required to work in participatory and collaborative environments. We must then design opportunities for learners to hone the skills of collective cognition and to work effectively in these environments. Since the Big Think strategies give learners practice with these skills consequently they become better and better at collaborative knowledge building and learning to learn.

In addition to a solid return on content acquired, the multi-layered Big Think is designed to help learners reflect on the processes used during the research process or unit of study and consider what worked, what did not, and why. This information is again pooled and examined for patterns and inconsistencies. Together strategies are developed to tackle problems and build on successes. Learners develop a new found efficacy and a positive mindset. They begin to see the importance of personal effort. They expect to get better because they have a plan (see Figure 1).

Sample Questions During a Content Big Think Activity

So What?
• What are the important ideas we explored?
• What does this tell us about the topic?
• What does this mean?
• What new understandings emerge?

What Next?
• What new questions do we have?
• How can we use what we know?
• What else do we want to explore?

Sample Questions During a Process Big Think Activity (21st century skills)

So What?
• What strategies did we use to learn?
• How did these strategies work for us?
• Which worked well or did not work well and for whom?

What Next?
• How can we use what we learned to do better next time?
• What will we do next?
• Where else can we apply what we now know and can do?

RETURN #2 TEACHING FOR LEARNING: WE REFLECT AS COACHES

Similarly, the adult teacher coaches need to conduct a Big Think at the end of the unit so

The Big Think activity consists of two elements that add up to increased knowledge building and real growth.

they know how to tweak their game plan for next time. Everyone involved in the collaborative venture—classroom teachers, teacher-librarians, teacher technologists, and other specialists need to put their heads together and debrief the effectiveness of the learning experience. They need to examine all the evidence available: planning notes, assessment data, student testimonials, reflections, visual documentation, and student products. They need to ask revealing and probing questions.

Sample Questions During a Coach's Content and Process Big Think Activity

So What?
- What did students learn? How did they learn it? Why is this important?
- What went well? What did not work? Why?
- Were all learners engaged?
- How well did differentiation strategies work?
- Does the assessment data give us a clear picture of student learning?
- Did the timing and chunking of the unit work?
- What learning environment problems did we encounter e.g. space, technologies, resources?
- How was understanding enhanced by the Big think?
- What process problems and successes were uncovered by the learners during their Big Think?

What Next?
- What new questions do we have?
- How can we use what we now know to do better next time?
- What actions should we take?

RETURN #3 SCHOOL IMPROVEMENT

Finally there is the opportunity to triple your investment. Reflective, informed learning, and teaching equals continuous growth—the foundation of sustained school improvement.

Teacher-librarians need to capture data from Big Think activities and include this information in data collection of school-wide achievement. It is an effective way to document value added by school library interventions. When we can demonstrate that two heads are better than one, when classroom teachers invest in working with teacher-librarians, the rewards are irresistible.

Too often learners are left out of the assessment piece. However, when students feel invested, they just might make greater strides toward excellence. When teachers are empowered with a process for improving their teaching and when they are supported and encouraged to adopt a strategic approach to teaching with learning in mind, then confidence and passion are restored. The Big Think creates this participatory culture where everyone is moving along toward a winning season; all are focused and confident that their goals are achievable.

We are at a turning point in education. Finances are limited, timetables and curriculum are overstuffed, and students and teachers are under pressure to perform. We have to achieve more in the same time with fewer resources. We must make every minute count. It really is the time to work smarter and to focus our efforts on strategies that ensure success and progress.

The 21st Century Skills movement has put further demands on education that must be addressed if we are to keep pace with global forces driving the need for a more elastic curriculum that will truly prepare learners for their world.

In our enthusiasm to prepare learners with evolving skills and literacies and equip them for learning in a shifting landscape, we must be careful not to shortchange content learning. It is not a matter of either or, but a thoughtful approach to the design of learning that carefully matches needed skills with desired content targets. In a recent article "21st Century Skills: The Challenges Ahead" in *Education Leadership*, authors Andrew Rotherham and Daniel Willingham (2009) state that, "the issue is how to meet the challenges of delivering content and skills in a rich way that genuinely improves outcomes for students" (p. 16).

The Think Models are the perfect teaching tools to assist educators in skillfully aligning 21st Century Skills with desired content in any discipline. The culminating Big Think is the final review or assessment piece that deepens and broadens understanding and develops collective knowledge as well as informing teachers and learners just where they are in terms of mastering skills and content. Teacher-librarians working with classroom teachers and other specialists can lead the way in ensuring both content and skills are valued in a 21st Century curriculum. The Big Think is consequently a vehicle for and a thermometer of whole school improvement. Process drives content and cannot be separated if real, long lasting, learning is to occur.

SO JUST WHAT IS A BIG THINK?

We propose that at the end of every learning experience educators invest a few minutes in a metacognitive exercise that will make learners more mindful of what they have gained in the way of knowledge, skills, and learning strategies. For the purpose of this article we will concentrate on the types of learning activities teacher-librarians most often are engaged in with learners: research and inquiry lessons and units based on content learning, as well as literature-based studies.

At the end of a typical unit, learners usually share their product or present their findings, get a grade, and move on to the next unit of study. Just when our students have enough knowledge about a topic to actually discuss it with some expertise, we slam the door shut on that topic and hope the individual learning will be retained. Occasionally we see evidence of individual self reflection but rarely collective cognition and synthesis of what we now know as a group.

Metacognition is basically the ability to reflect on an experience and reason about what worked and what did not, and why, and then strategize for improvement. Thus metacognition is critical to learning how to learn. Without an opportunity to think about learning, students rarely unpack the importance of new knowledge gained or make connections to bigger ideas and concepts. They certainly will not grow as learners without opportunity to analyze their strengths and weaknesses and set goals for improvement.

When a unit of study is completed learners are then ready to play the game of learning. Each individual has something special to bring to the field. We design a Big Think experience to capitalize on learning from the main event and ask learners to do some deep thinking about the content in order to build personal and collective

The Big Think Changes Everything
Nine Metacognitive Strategies that Make the Unit End Just the Beginning of Learning

STRATEGY Teachers and learners think about content and process	WHAT? The information to knowledge journey	WHY? Knowledge building and real growth	HOW? Make connections as a group between what I know and what we discovered. Develop what we now know.
Active Discussion	Small and large group face to face and/or virtual discussion ignited by a question or scenario	To develop, clarify, interpret, empathize, defend, understand	Informal discussion, formal panel, debate, press conference, blog, wiki, interactive video conferencing, etc.
Create New Questions	Collaborative reflection, analysis, discovery, exploration of opinions and points of view directed by student developed questions	To create a culture of inquiry; to ensure personal relevance, perspective, purpose and direction for thinking, springboards for further actions, research, critical analysis	Use question building assists; question storming, Bloom's Taxonomy, De Bono's Thinking Hats, question matrix, etc.
Higher Order Thinking	Collaborative critical and creative thinking	To raise level of understanding, solve, infer, predict, evaluate, argue, innovate	Stretching, comparing, speculating, predicting, discovering effect and impact, analyzing, synthesizing, evaluating
Interact with an Expert	Confirm, amend, or enhance understandings, explore ideas and interpretations	To exchange ideas, glean new knowledge, gain perspective, add relevance, make real world connections	Interview, consultation, face to face and/or by videoconference, blog, Twitter, Skype, email; Real or virtual field trip, tour
New Problem or Challenge	Stimulate creative collaboration by presenting a new problem or challenge that draws on collective knowledge and expertise	Transfer and apply knowledge, solve problems, develop fluency and flexibility, simulate real life situations, make learning relevant	Introduce an element shift or what if scenario, problems possibilities jigsaw, concept jigsaw, teach or coach
Thoughtful Writing	Construct and articulate deep understanding through a process of collaborative writing	Consider alternate ideas and perspectives, construct meaning, write collaboratively, stimulate curiosity and interdependent thinking	Concept writing, quick write, chart, letter, wish list, zine, wikis and other Web 2.0 tools
Construct Visuals	Active building of knowledge through visual representations	To clarify concepts, build knowledge, convey meaning on sight, accommodate visual learners, enable those with language or learning deficiencies	Charts, graphs, flow charts, timelines, webs, illustrations, cartoons, comic strips, concept mapping software, and other technology applications
ReCreate	Transform information and ideas to a new medium	To present information and ideas via a new medium, build understanding of concepts and events, tap into emotional intelligence, develop empathy	Create a skit, dramatic representation, collage, web, video, game, podcast, and other creative technology applications
Sandbox	Play with ideas and information to create or invent something new	Brain based learning, utilizing all senses, stimulates curiosity, wonder and discovery, ownership and freedom of choice, ignite renewed passion for learning	Creative technology applications, music, drama, visual arts, video, tangible manipulatives

Table 1. Loertscher, D. V., Koechlin, C., & Zwaan, S. *The Big Think: 9 Metacognitive Strategies That Make the Unit End Just the Beginning of Learning.*

knowledge. We know from brain-based research that long term memory hinges on making connections and processing information in many different ways. The Big Think strategies apply many principles of brain-based learning and thus contribute to real long lasting learning.

Another foundational goal of the Big Think is for learners to improve skills, develop habits of mind, and gain responsibilities conducive to learning how to learn. Carol Dweck (2006) refers to this needed ability as a growth mindset, in her book the *New Psychology of Success*. Dweck tells us that given a Growth Mindset, necessary resources, opportunity, and the transformative power of effort, we can in fact reach our full potential. We can study and apply the mindset psychology in our efforts to improve outcomes for learners and help them become more self reliant. With greater student and staff involvement in assessment we can demonstrate the value of effort. When we work as teams we can provide opportunities to make the learning experiences in our schools exemplary. We can assist in establishing the habit of personal and professional growth, reflective practice, personal responsibility, and confidence.

Implementing Big Think

During the Big Think it is critical for teachers to still be involved and provide needed guidance and feedback if learners are to get better. The nine metacognitive strategies provide learners practice with a variety of learning how to learn skills, but as Rotherham and Willingham (2009) also point out in their article, "Experience means only that you use a skill; practice means that you try to improve by noticing what you are doing wrong and formulating strategies to do better. Practice also requires feedback, usually from someone more skilled than you are" (p. 18). Metacognition and useful feedback becomes part of the culture or game plan of learning in our schools and everyone, teachers and students, get better and better.

We have developed these nine basic strategies to provide the best potential for engagement and high think. The Big Think activities do not need to be time consuming. They can take anywhere from five minutes to a class period or longer in the event that more involved What Next activities are sparked. The point is that the Big Think needs to be designed as part of the lesson or unit because it is just too important to neglect. (Table 1 provides an overview of each strategy).

BACK TO THE GAME PLAN

We call on teacher-librarians to coach their staff and students on the many benefits of Big Think strategies. At the end of a unit keep the thinking flowing and strive for deeper understandings, facilitate transformations of learning, and spark new student innovations and creations. Invest in the design of Big Think activities to help learners become more mindful of what they are learning, how they are learning it, and why; help teachers become reflective practitioners; and contribute to whole school improvement and excellence.

This is the winning formula!

• Collaborate with classroom teachers and other specialists to design and teach research and inquiry units using the Think Models. Culminate with a Big Think of content and processes to further elevate library projects so that the product or presentation is no longer the end; it is just the beginning of real learning!

• Conduct a Big Think with teaching partners.

• Share evidence with the entire school community.

• Reflect, react, and realize improved learning.

FURTHER READING

Teaching with the Brain in Mind, 2nd edition. Eric Jensen. ASCD, 2005.

Building Info Smarts: How to work with all kinds of information and make it your own. Carol Koechlin and Sandi Zwaan. Pembroke, 2008.

Q Tasks: How to Empower Students to Ask Questions and Care About Answers. Carol Koechlin and Sandi Zwaan. Pembroke, 2006.

The New Learning Commons: Where Learners Win. David Loertscher, Carol Koechlin, and Sandi Zwaan. Hi Willow Research and Publishing, 2008.

Brain Friendly School Libraries. Judith Sykes. Libraries Unlimited, 2006.

Brain Matters: Translating Research into Classroom Practice. Patricia Wolfe. ASCD, 2001.

REFERENCES

Dweck, C. S. *Mindset: The new psychology of success.* NY: Ballantine.

Loertscher, D. V., Koechlin, C., & Zwaan, S. *The Big Think: 9 metacognitive strategies that make the unit end just the beginning of learning.* Salt Lake City, UT: Hi Willow Research and Publishing.

Loertscher, D. V., Koechlin, C., Zwaan, S. (2007). *Beyond bird units.* Salt Lake City, UT: Hi Willow Research and Publishing.

Rotherhamm, A. J. & Willingham, D. (2009). 21st century skills: The challenges ahead. *Education Leadership* (67)1, 16-21.

Carol Koechlin and **Sandi Zwaan** have worked as classroom teachers, teacher-librarians, educational consultants, staff development leaders, and instructors for Educational Librarianship courses for York University and University of Toronto. In their quest to provide teachers with strategies to make learning opportunities more meaningful, more reflective, and more successful, they have led staff development sessions for teachers in both Canada and the United States. They continue to contribute to the field of information literacy and school librarianship by coauthoring a number of books and articles for professional journals. Their work has been recognized both nationally and internationally and translated into French, German, Italian, and Chinese. They may be contacted at *koechlin@sympatico.ca* and *sandi.zwaan@sympatico.ca*.

Author, Title, and Subject Index

Note: Page numbers in this index refer to the beginning of the article authored by, article begins, or a theme or subject of the article.

Achieving Teaching and Learning Excellence with Technology, 85

Administrators, 13, 20, 27

Advanced Contemporary Literacy: An Integrated Approach to Reading, 124

Alevy, Jennifer, 137

Allen Centre, 27

Assessment, 31, 37, 55, 81d, 85, 111, 117, 129, 137

Bentheim, Christina A., 31

Big Think: Reflecting, Reacting, and Realizing Improved Learning. 139

Byrne, Richard, 98

Cabrera, Derek, 43

Calamari, 101

Carroll, Greg, 27

Chelmsford High School, 13

Cicchetti, Robin, 20

Collaboration, 81a, 111, 122

Colosi, Laura, 43

Concord-Carlisle High School, 20

Concord-Carlisle Transitions to a Learning Commons, 20

Cooper-Simon, Sheila, 81d

Creating Personal Learning Through Self-Assessment, 129

Creative thinking, 66

Creativity, 66, 120

Critical thinking, 43, 73, 139

Cultivating Curious Minds: Teaching for Innovation Through Open-Inquiry Learning, 66

Curriculum, 124

Curriculum and the Learning Commons, 37

Curriculum research, 55

Curriculum, the Library/Learning Commons and Teacher Librarians: Myths and Realities in the Second Decade, 37

Czarnecki, Kelly, 120

D.L. "Dusty" Dickens Elementary School Learning Commons, 31

Data collection 31, 37, 55, 111, 129, 137

Davis, Vicki, 55

Derry, Bill, 117

Differentiation strategies, 60, 81a

Diggs, Valerie, 13

Donham, Jean, 129

Effect of Web 2.0 on Teaching and Learning, 98

Elementary school Learning Commons, 27, 31

English learners, 122

Everyone Wins: Differentiation in the School Library, 60

Facebook, 104

Facilities, 13, 20

Fodeman, Doug, 104

From Book Museum to Learning Commons: Riding the Transformation Train, 31

From Library to Learning Commons: a Metamorphosis by Valerie Diggs with Editorial Comments, 13

Gifted Readers and Libraries: A Natrual Fit, 77

Gifted students, 77

Global Education, 55

Google Apps, 94

Harada, Violet H., 111

Haslam-Odoardi, Reecca, 77

High School Learning

Influencing Positive Change: The Vital Behaviors to Turn Schools Toward Success, 55

Impact of Facebook on our Students, 104Information and Technology Literacy, 117

Information literacy, 73, 117, 129

Information Literate? Just Turn the Children Loose!, 73

Inquiry, 66, 73, 111, 129

Knodt, Jean Sausele, 66

Koechlin, Carol, 3, 60, 139

Lamb, Christopher, 122

Leadership, 13, 20, 37, 111, 117, 120, 124

Learning Commons, Foundational ideas, 3

Learning Commons is Alive and Well in New Zealand, 27

Librarians and Learning Specialists: Moving From the Margins to the Mainstream of School Leadership, 111

Library is the Place: Knowledge and Thinking, Thinking and Knowledge, 43

Loertscher, David V., 13, 37, 49, 85

Lopez, Carol, 122

Manipulatives, 43

Marcoux, Elizabeth "Betty," 49, 85

Metacognition, 139

Monroe, Marje, 104

Mounter, Joy, 73

Nevin, Roger, 81a, 94

New Zealand, 27

Open inquiry, 66

Our Instruction Does Matter: Data Collected from Students' Works Cited Speaks Volumes by, 137

Pointer, Sarah, 137

Porter, Winnie, 122

Privacy, 104

Professional development, 85, 117, 124

Program transformation, 13, 20

Public and school library cooperation, 120

Reading, 49, 77, 122

Reading research, 49

Rethinking collaboration: transforming Web 2.0 thinking into real time behavior, 81d

Role of the School Library in the Reading Program, 49

Sargeant, Cynthia, 81a

School and public library cooperation, 120

Security, 104

Social networking, 81d, 104, 129

Stedman, Peggy, 27

Supporting 21st Century Learning Through Google Apps, 94

Swarner, Sharon, 124

Teacher librarian, role of, 49, 111

Technology, 55, 81d, 85, 94, 98, 101, 104, 117

Technology Leadership: Kelly Czarnecki, 120

Time is Now: Transform Your School Library into a Learning Commons, 3

Three Heads and Better Than One: The Reading Coach, The Classroom Teacher, and the Teacher Librarian, 122

Using the library learning commons to reengage disengaged students and making it a student-friendly place, 81a

Westwood Schools, Camilla, GA, 55

Wireless, 101

WLANS for the 21st Century Library, 101

Zmuda, Alison, 111

Zwaan, Sandi, 3, 60,